T0005497

ALTERED
TRAITS

ALTERED TRAITS

Science Reveals How Meditation Changes Your Mind, Brain, and Body

DANIEL
GOLEMAN

AND

RICHARD J.
DAVIDSON

AVERY   an imprint of Penguin Random House   New York

AVERY

An imprint of Penguin Random House LLC
375 Hudson Street
New York, New York 10014

First trade paperback edition 2018
Copyright © 2017 by Daniel Goleman and Richard J. Davidson
Penguin supports copyright. Copyright fuels creativity, encourages diverse
voices, promotes free speech, and creates a vibrant culture. Thank you for buying
an authorized edition of this book and for complying with copyright laws by
not reproducing, scanning, or distributing any part of it in any form without
permission. You are supporting writers and allowing Penguin to continue
to publish books for every reader.

Most Avery books are available at special quantity discounts for bulk purchase
for sales promotions, premiums, fund-raising, and educational needs. Special
books or book excerpts also can be created to fit specific needs. For details,
write SpecialMarkets@penguinrandomhouse.com.

Paperback edition ISBN 9780399184390
Hardcover edition ISBN 9780399184383
Export edition ISBN 9780735220317

Printed in the United States of America
11th Printing

Book design by Ellen Cipriano

CONTENTS

The Deep Path and the Wide

One bright fall morning, Steve Z, a lieutenant colonel working in the Pentagon, heard a "crazy, loud noise," and instantly was covered in debris as the ceiling caved in, knocking him to the floor, unconscious. It was September 11, 2001, and a passenger jet had smashed into the huge building, very near to Steve's office.

The debris that buried Steve saved his life as the plane's fuselage exploded, a fireball of flames scouring the open office. Despite a concussion, Steve returned to work four days later, laboring through feverish nights, 6:00 p.m. to 6:00 a.m., because those were daytime hours in Afghanistan. Soon after, he volunteered for a year in Iraq.

"I mainly went to Iraq because I couldn't walk around the Mall without being hypervigilant, wary of how people looked at me, totally on guard," Steve recalls. "I couldn't get on an elevator, I felt trapped in my car in traffic."

His symptoms were classic post-traumatic stress disorder. Then came the day he realized he couldn't handle this on his own. Steve ended up with a psychotherapist he still sees. She led him, very gently, to try mindfulness.

Mindfulness, he recalls, "gave me something I could do to help feel more calm, less stressed, not be so reactive." As he practiced more, added loving-kindness to the mix, and went on retreats, his PTSD symptoms gradually became less frequent, less intense. Although his irritability and restlessness still came, he could see them coming.

Tales like Steve's offer encouraging news about meditation. We have been meditators all our adult lives, and, like Steve, know for ourselves that the practice has countless benefits.

But our scientific backgrounds give us pause, too. Not everything chalked up to meditation's magic actually stands up to rigorous tests. And so we have set out to make clear what works and what does not.

Some of what you know about meditation may be wrong. But what is true about meditation you may not know.

Take Steve's story. The tale has been repeated in endless variations by countless others who claim to have found relief in meditation methods like mindfulness—not just from PTSD but from virtually the entire range of emotional disorders.

Yet mindfulness, part of an ancient meditation tradition, was not intended to be such a cure; this method was only recently adapted as a balm for our modern forms of angst. The original aim, embraced in some circles to this day, focuses on a deep exploration of the mind toward a profound alteration of our very being.

On the other hand, the pragmatic applications of meditation— like the mindfulness that helped Steve recover from trauma—appeal widely but do not go so deep. Because this wide approach has easy access, multitudes have found a way to include at least a bit of meditation in their day.

There are, then, two paths: the deep and the wide. Those two paths are often confused with each other, though they differ greatly.

We see the deep path embodied at two levels: in a pure form, for example, in the ancient lineages of Theravada Buddhism as practiced in Southeast Asia, or among Tibetan yogis (for whom we'll see some remarkable data in chapter eleven, "A Yogi's Brain"). We'll call this most intensive type of practice Level 1.

At Level 2, these traditions have been removed from being part of a total lifestyle—monk or yogi, for example—and adapted into forms more palatable for the West. At Level 2, meditation comes in forms that leave behind parts of the original Asian source that might not make the cross-cultural journey so easily.

Then there are the wide approaches. At Level 3, a further remove takes these same meditation practices out of their spiritual context and distributes them ever more widely—as is the case with mindfulness-based stress reduction (better known as MBSR), founded by our good friend Jon Kabat-Zinn and taught now in thousands of clinics and medical centers, and far beyond. Or Transcendental Meditation (TM), which offers classic Sanskrit mantras to the modern world in a user-friendly format.

The even more widely accessible forms of meditation at Level 4 are, of necessity, the most watered-down, all the better to render them handy for the largest number of people. The current vogues of mindfulness-at-your-desk, or via minutes-long meditation apps, exemplify this level.

We foresee also a Level 5, one that exists now only in bits and pieces, but which may well increase in number and reach with time. At

Level 5, the lessons scientists have learned in studying all the other levels will lead to innovations and adaptations that can be of widest benefit—a potential we explore in the final chapter, "A Healthy Mind."

The deep transformations of Level 1 fascinated us when we originally encountered meditation. Dan studied ancient texts and practiced the methods they describe, particularly during the two years he lived in India and Sri Lanka in his grad school days and just afterward. Richie (as everyone calls him) followed Dan to Asia for a lengthy visit, likewise practicing on retreat there, meeting with meditation scholars—and more recently has scanned the brains of Olympic-level meditators in his lab at the University of Wisconsin.

Our own meditation practice has been mainly at Level 2. But from the start, the wide path, Levels 3 and 4, has also been important to us. Our Asian teachers said if any aspect of meditation could help alleviate suffering, it should be offered to all, not just those on a spiritual search. Our doctoral dissertations applied that advice by studying ways meditation could have cognitive and emotional payoffs.

The story we tell here mirrors our own personal and professional journey. We have been close friends and collaborators on the science of meditation since the 1970s, when we met at Harvard during graduate school, and we have both been practitioners of this inner art over all these years (although we are nowhere near mastery).

While we were both trained as psychologists, we bring complementary skills to telling this story. Dan is a seasoned science journalist who wrote for the *New York Times* for more than a decade. Richie, a neuroscientist, founded and heads the University of Wisconsin's Center for Healthy Minds, in addition to directing the brain imaging laboratory at the Waisman Center there, replete with its own fMRI, PET scanner, and a battery of cutting-edge data analysis programs, along with hundreds of servers for the heavy-duty computing required for this work. His research group numbers more than a hundred experts, who range from physicists, statisticians, and computer scientists to neuroscientists and psychologists, as well as scholars of meditative traditions.

Coauthoring a book can be awkward. We've had some of that, to be sure—but whatever drawbacks coauthorship brought us has been vastly overshadowed by the sheer delight we find in working together. We've been best friends for decades but labored separately over most of our careers. This book has brought us together again, always a joy.

You are holding the book we had always wanted to write but could not. The science and the data we needed to support our ideas have only recently matured. Now that both have reached a critical mass, we are delighted to share this.

Our joy also comes from our sense of a shared, meaningful mission: we aim to shift the conversation with a radical reinterpretation of what the actual benefits of meditation are—and are not—and what the true aim of practice has always been.

THE DEEP PATH

After his return from India in the fall of 1974, Richie was in a seminar on psychopathology back at Harvard. Richie, with long hair and attire in keeping with the zeitgeist of Cambridge in those times—including a colorful woven sash that he wore as a belt—was startled when his professor said, "One clue to schizophrenia is the bizarre way a person dresses," giving Richie a meaningful glance.

And when Richie told one of his Harvard professors that he

wanted to focus his dissertation on meditation, the blunt response came immediately: that would be a career-ending move.

Dan set out to research the impacts of meditation that uses a mantra. On hearing this, one of his clinical psychology professors asked with suspicion, "How is a mantra any different from my obsessive patients who can't stop saying 'shit-shit-shit'?"[1] The explanation that the expletives are involuntary in the psychopathology, while the silent mantra repetition is a voluntary and intentional focusing device, did little to placate him.

These reactions were typical of the opposition we faced from our department heads, who were still responding with knee-jerk negativity toward anything to do with consciousness—perhaps a mild form of PTSD after the notorious debacle involving Timothy Leary and Richard Alpert. Leary and Alpert had been very publicly ousted from our department in a brouhaha over letting Harvard undergrads experiment with psychedelics. This was some five years before we arrived, but the echoes lingered.

Despite our academic mentors' seeing our meditation research as a blind alley, our hearts told us this was of compelling import. We had a big idea: beyond the pleasant states meditation can produce, the real payoffs are the lasting *traits* that can result.

An altered trait—a new characteristic that arises from a meditation practice—endures apart from meditation itself. Altered traits shape how we behave in our daily lives, not just during or immediately after we meditate.

The concept of altered traits has been a lifelong pursuit, each of us playing synergistic roles in the unfolding of this story. There were Dan's years in India as an early participant-observer in the Asian roots of these mind-altering methods. And on Dan's return to America he was a not-so-successful transmitter to contemporary psychology of beneficial changes from meditation and the ancient working models for achieving them.

Richie's own experiences with meditation led to decades pursuing the science that supports our theory of altered traits. His research group has now generated the data that lend credence to what could otherwise seem mere fanciful tales. And by leading the creation of a fledgling research field, contemplative neuroscience, he has been grooming a coming generation of scientists whose work builds on and adds to this evidence.

In the wake of the tsunami of excitement over the wide path, the alternate route so often gets missed: that is, the deep path, which has always been the true goal of meditation. As we see it, the most compelling impacts of meditation are not better health or sharper business performance but, rather, a further reach toward our better nature.

A stream of findings from the deep path markedly boosts science's models of the upper limits of our positive potential. The further reaches of the deep path cultivate enduring qualities like selflessness, equanimity, a loving presence, and impartial compassion—highly positive altered traits.

When we began, this seemed big news for modern psychology—if it would listen. Admittedly, at first the concept of altered traits had scant backing save for the gut feelings we had from meeting highly seasoned practitioners in Asia, the claims of ancient meditation texts, and our own fledgling tries at this inner art. Now, after decades of silence and disregard, the last few years have seen ample findings that bear out our early hunch. Only of late have the scientific data reached critical mass, confirming what our intuition and the texts told us: these deep changes are external signs of strikingly different brain function.

Much of that data comes from Richie's lab, the only scientific center that has gathered findings on dozens of contemplative masters, mainly Tibetan yogis—the largest pool of deep practitioners studied anywhere.

These unlikely research partners have been crucial in building a scientific case for the existence of a way of being that has eluded modern thought, though it was hiding in plain sight as a goal of the world's major spiritual traditions. Now we can share scientific confirmation of these profound alterations of being—a transformation that dramatically ups the limits on psychological science's ideas of human possibility.

The very idea of "awakening"—the goal of the deep path—seems a quaint fairy tale to a modern sensibility. Yet data from Richie's lab, some just being published in journals as this book goes to press, confirm that remarkable, positive alterations in brain and behavior along the lines of those long described for the deep path are not a myth but a reality.

THE WIDE PATH

We have both been longtime board members of the Mind and Life Institute, formed initially to create intensive dialogues between the Dalai Lama and scientists on wide-ranging topics.[2] In 2000 we organized one on "destructive emotions," with several top experts on emotions, including Richie.[3] Midway through that dialogue the Dalai Lama, turning to Richie, made a provocative challenge.

His own tradition, the Dalai Lama observed, had a wide array of time-tested practices for taming destructive emotions. So, he urged, take these methods into the laboratory in forms freed from religious trappings, test them rigorously, and if they can help people lessen their destructive emotions, then spread them widely to all who might benefit.

That fired us up. Over dinner that night—and several nights following—we began to plot the general course of the research we report in this book.

The Dalai Lama's challenge led Richie to refocus the formidable power of his lab to assess both the deep and the wide paths. And, as founding director of the Center for Healthy Minds, Richie has spurred work on useful, evidence-based applications suitable for schools, clinics, businesses, even for cops—for anyone, anywhere, ranging from a kindness program for preschoolers to treatments for veterans with PTSD.

The Dalai Lama's urging catalyzed studies that support the wide path in scientific terms, a vernacular welcomed around the globe. Meanwhile the wide way has gone viral, becoming the stuff of blogs, tweets, and snappy apps. For instance, as we write this, a wave of enthusiasm surrounds mindfulness, and hundreds of thousands—maybe millions—now practice the method.

But viewing mindfulness (or any variety of meditation) through a scientific lens starts with questions like: When does it work, and when does it not? Will this method help everyone? Are its benefits any different from, say, exercise? These are among the questions that brought us to write this book.

Meditation is a catch-all word for myriad varieties of contemplative practice, just as *sports* refers to a wide range of athletic activities. For both sports and meditation, the end results vary depending on what you actually do.

Some practical advice: for those about to start a meditation practice, or who have been grazing among several, keep in mind that as with gaining skill in a given sport, finding a meditation practice that appeals to you and sticking with it will have the greatest benefits. Just find one to try, decide on the amount of time each day you can realistically practice daily—even as short as a few minutes—try it for a month, and see how you feel after those thirty days.

Just as regular workouts give you better physical fitness, most any type of meditation will enhance mental fitness to some degree. As we'll see, the specific benefits from one or another type get stronger the more total hours of practice you put in.

A CAUTIONARY TALE

Swami X, as we'll call him, was at the tip of the wave of meditation teachers from Asia who swarmed to America in the mid-1970s, during our Harvard days. The swami reached out to us saying he was eager to have his yogic prowess studied by scientists at Harvard who could confirm his remarkable abilities.

It was the height of excitement about a then new technology, biofeedback, which fed people instant information about their physiology—blood pressure, for instance—which otherwise was beyond their conscious control. With that new incoming signal, people were able to nudge their body's operations in healthier directions. Swami X claimed he had such control without the need for feedback.

Happy to stumble on a seemingly accomplished subject for research, we were able to finagle the use of a physiology lab at Harvard Medical School's Massachusetts Mental Health Center.[4]

But come the day of testing the swami's prowess, when we asked him to lower his blood pressure, he raised it. When asked to raise it, he lowered it. And when we told him this, the swami berated us for serving him "toxic tea" that supposedly sabotaged his gifts.

Our physiological tracings revealed he could do none of the mental feats he had boasted about. He did, however, manage to put his heart into atrial fibrillation—a high-risk biotalent—with a method he called "dog samadhi," a name that mystifies us to this day.

From time to time the swami disappeared into the men's room to smoke a *bidi* (these cheap cigarettes, a few flakes of tobacco wrapped in a plant leaf, are popular throughout India). A telegram from friends in India soon after revealed that the "swami" was actually the former manager of a shoe factory who had abandoned his wife and two children and come to America to make his fortune.

No doubt Swami X was seeking a marketing edge to attract disciples. In his subsequent appearances he made sure to mention that "scientists at Harvard" had studied his meditative prowess. This was an early harbinger of what has become a bountiful harvest of data refried into sales hype.

With such cautionary incidents in mind, we bring open but skeptical minds—the scientist's mind-set—to the current wave of meditation research. For the most part we view with satisfaction the rise of the mindfulness movement and its rapidly growing reach in schools, business, and our private lives—the wide approach. But we bemoan how the data all too often is distorted or exaggerated when science gets used as a sales hook.

The mix of meditation and monetizing has a sorry track record as a recipe for hucksterism, disappointment, even scandal. All too often, gross misrepresentations, questionable claims, or distortions of scientific studies are used to sell meditation. A business website, for instance, features a blog post called "How Mindfulness Fixes Your

Brain, Reduces Stress, and Boosts Performance." Are these claims justified by solid scientific findings? Yes and no—though the "no" too easily gets overlooked.

Among the iffy findings gone viral with enthusiastic claims: that meditation thickens the brain's executive center, the prefrontal cortex, while shrinking the amygdala, the trigger for our freeze-fight-or-flight response; that meditation shifts our brain's set point for emotions into a more positive range; that meditation slows aging; and that meditation can be used to treat diseases ranging from diabetes to attention deficit hyperactivity disorder.

On closer look, each of the studies on which these claims are based has problems with the methods used; they need more testing and corroboration to make firm claims. Such findings may well stand up to further scrutiny—or maybe not.

The research reporting amygdala shrinkage, for instance, used a method to estimate amygdala volume that may not be very accurate. And one widely cited study describing slower aging used a very complex treatment that included some meditation but was mixed with a special diet and intensive exercise as well; the impact of meditation per se was impossible to decipher.

Still, social media are rife with such claims—and hyperbolic ad copy can be enticing. So we offer a clear-eyed view based on hard science, sifting out results that are not nearly as compelling as the claims made for them.

Even well-meaning proponents have little guidance in distinguishing between what's sound and what's questionable—or just sheer nonsense. Given the rising tide of enthusiasm, our more sober-minded take comes not a moment too soon.

A note to readers. The first three chapters cover our initial forays into meditation, and the scientific hunch that motivated our quest. Chapters four through twelve narrate the scientific journey, with each chapter devoted to a particular topic like attention or compassion; each of these has an "In a Nutshell" summary at the end for those who are more interested in what we found than how we got there. In chapters eleven and twelve we arrive at our long-sought destination, sharing the remarkable findings on the most advanced meditators ever studied. In chapter thirteen, "Altering Traits," we lay out the benefits of meditation at three levels: beginner, long-term, and "Olympic." In our final chapter we speculate on what the future might bring, and how these findings might be of greater benefit not just to each of us individually but to society.

THE ACCELERATION

As early as the 1830s, Thoreau and Emerson, along with their fellow American Transcendentalists, flirted with these Eastern inner arts. They were spurred by the first English-language translations of ancient spiritual texts from Asia—but had no instruction in the practices that supported those texts. Almost a century later, Sigmund Freud advised psychoanalysts to adopt an "even-hovering attention" while listening to their clients—but again, offered no method.

The West's more serious engagement took hold mere decades ago, as teachers from the East arrived, and as a generation of Westerners traveled to study meditation in Asia, some returning as teachers. These forays paved the way for the current acceleration of the wide path, along with fresh possibilities for those few who choose to pursue the deep way.

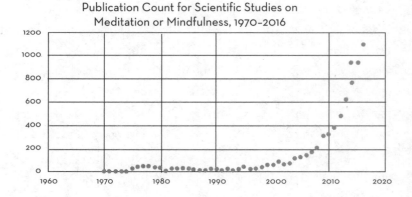

Publication Count for Scientific Studies on
Meditation or Mindfulness, 1970–2016

In the 1970s, when we began publishing our research on meditation, there were just a handful of scientific articles on the topic. At last count there numbered 6,838 such articles, with a notable acceleration of late. For 2014 the annual number was 925, in 2015 the total was 1,098, and in 2016 there were 1,113 such publications in the English language scientific literature.[5]

PRIMING THE FIELD

It was April 2001, on the top floor of the Fluno Center on the campus of the University of Wisconsin–Madison, and we were convening with the Dalai Lama for an afternoon of scientific dialogue on meditation research findings. Missing from the room was Francisco Varela, a Chilean-born neuroscientist and head of a cognitive neuroscience laboratory at the French National Center for Scientific Research in Paris. His remarkable career included cofounding the Mind and Life Institute, which had organized this very gathering.

As a serious meditation practitioner, Francisco could see the promise for a full collaboration between seasoned meditators and the scientists studying them. That model became standard practice in Richie's lab, as well as others.

Francisco had been scheduled to participate, but he was fighting liver cancer and a severe downturn meant he could not travel. He was in his bed at home in Paris, close to dying.

This was in the days before Skype and videoconferencing, but Richie's group managed a two-way video hookup between our meeting room and Francisco's bedroom in his Paris apartment. The Dalai Lama addressed him very directly, looking closely into the camera. They both knew that this would be the very last time they would see each other in this lifetime.

The Dalai Lama thanked Francisco for all he had done for science and for the greater good, told him to be strong, and said that they would remain connected forever. Richie and many others in the room had tears streaming down, appreciating the momentous import of the moment. Just days after the meeting, Francisco passed away.

Three years later, in 2004, an event occurred that made real a dream Francisco had often talked about. At the Garrison Institute, an hour up the Hudson River from New York City, one hundred scientists, graduate students, and postdocs had gathered for the first in what has become a yearly series of events, the Summer Research Institute (SRI), a gathering devoted to furthering the rigorous study of meditation.

The meetings are organized by the Mind and Life Institute, itself formed in 1987 by the Dalai Lama, Francisco, and Adam Engle, a lawyer turned businessman. We were founding board members. The mission of Mind and Life is "to alleviate suffering and promote flourishing by integrating science with contemplative practice."

Mind and Life's summer institute, we felt, could offer a more welcoming reality for those who, like us in our grad school days, wanted to do research on meditation. While we had been isolated pioneers, we wanted to knit together a community of like-minded scholars and scientists who shared this quest. They could be supportive of each other's work at a distance, even if they were alone in their interests at their own institution.

Details of the SRI were hatched over the kitchen table in Richie's home in Madison, in a conversation with Adam Engle. Richie and a handful of scientists and scholars then organized the first summer program and served as faculty for the week, featuring topics like the cognitive neuroscience of attention and mental imagery. As of this writing, thirteen more meetings have followed (with two so far in Europe, and possibly future meetings in Asia and South America).

Beginning with the very first SRI, the Mind and Life Institute began a program of small grants named in honor of Francisco. These few dozen, very modest Varela research awards (up to $25,000, though most research of this kind takes far more in funding) have leveraged more than $60 million in follow-on funding from foundations and US federal granting agencies. And the initiative has borne plentiful fruit: fifty or so graduates of the SRI have published several hundred papers on meditation.

As these young scientists entered academic posts, they swelled the numbers of researchers doing such studies. They have driven in no small part the ever-growing numbers of scientific studies on meditation.

At the same time, more established scientists have shifted their focus toward this area as results showed valuable yield. The findings rolling out of Richie's brain lab at the University of Wisconsin—and labs of other scientists, from the medical schools of Stanford and Emory, Yale and Harvard, and far beyond—routinely make headlines.

Given meditation's booming popularity, we feel a need for a hard-nosed look. The neural and biological benefits best documented by sound science are not necessarily the ones we hear about in the press, on Facebook, or from email marketing blasts. And some of those trumpeted far and wide have little scientific merit.

Many reports boil down to the ways a short daily dose of meditation alters our biology and emotional life for the better. This news, gone viral, has drawn millions worldwide to find a slot in their daily routine for meditation.

But there are far greater possibilities—and some perils. The moment has come to tell the bigger tale the headlines are missing.

There are several threads in the tapestry we weave here. One can be seen in the story of our decades-long friendship and our shared sense of a greater purpose, at first a distant and unlikely goal but one in which we persisted despite obstacles. Another traces the emergence of neuroscience's evidence that our experiences shape our brains, a platform supporting our theory that as meditation trains the mind, it reshapes the brain. Then there's the flood of data we've mined to show the gradient of this change.

At the outset, mere minutes a day of practice have surprising benefits (though not all those that are claimed). Beyond such payoffs at the beginning, we can now show that the more hours you practice, the greater the benefits you reap. And at the highest levels of practice we find true altered traits—changes in the brain that science has never observed before, but which we proposed decades ago.

2

Ancient Clues

Our story starts one early November morning in 1970, when the spire of the stupa in Bodh Gaya was lost to view, enveloped in the ethereal mist rising from the Niranjan River nearby. Next to the stupa stood a descendant of the very Bodhi Tree under which, legend has it, Buddha sat in meditation as he became enlightened.

Through the mist that morning, Dan glimpsed an elderly Tibetan monk amble by as he made his postdawn rounds, circumambulating the holy site. With short-cropped gray hair and eyeglasses as thick as the bottoms of Coke bottles, he fingered his mala beads while mumbling softly a mantra praising the Buddha as a sage, or *muni* in Sanskrit: "*Muni, muni, mahamuni, mahamuniya swaha!*"

A few days later, friends happened to bring Dan to visit that very monk, Khunu Lama. He inhabited a sparse, unheated cell, its concrete walls radiating the late-fall chill. A wooden-plank *tucket* served as both bed and day couch, with a small stand alongside for perching texts to read—and little else. As befits a monk, the room was empty of any private belongings.

From the early-morning hours until late into the night, Khunu Lama would sit on that bed, a text always open in front of him. Whenever a visitor would pop in—and in the Tibetan world that could be at just about any time—he would invariably welcome them with a kindly gaze and warm words.

Khunu's qualities—a loving attention to whoever came to see him, an ease of being, and a gentle presence—struck Dan as quite unlike, and far more positive than, the personality traits he had been studying for his degree in clinical psychology at Harvard. That training focused on negatives: neurotic patterns, overpowering burdensome feelings, and outright psychopathology.

Khunu, on the other hand, quietly exuded the better side of human nature. His humility, for instance, was fabled. The story goes that the abbot of the monastery, in recognition of Khunu's spiritual status, offered him as living quarters a suite of rooms on the monastery's top floor, with a monk to serve as an attendant. Khunu declined, preferring the simplicity of his small, bare monk's cell.

Khunu Lama was one of those rare masters revered by all schools of Tibetan practice. Even the Dalai Lama sought him out for teachings, receiving instructions on Shantideva's *Bodhicharyavatara*, a guide to the compassion-filled life of a bodhisattva. To this day, whenever the Dalai Lama teaches this text, one of his favorites, he credits Khunu as his mentor on the topic.

Before meeting Khunu Lama, Dan had spent months with an Indian yogi, Neem Karoli Baba, who had drawn him to India in the first place. Neem Karoli, known by the honorific Maharaji, was newly famous in the West as the guru of Ram Dass, who in those years toured the country with mesmerizing accounts of his transformation from Richard Alpert (the Harvard professor fired for experimenting with psychedelics, along with his colleague Timothy Leary) to a devotee of this old yogi. By accident, during Christmas break from his Harvard classes in 1968, Dan met Ram Dass, who had just returned from being with Neem Karoli in India, and that encounter eventually propelled Dan's journey to India.

Dan managed to get a Harvard Predoctoral Traveling Fellowship to India in fall 1970, and located Neem Karoli Baba at a small ashram in the Himalayan foothills. Living the life of a sadhu, Maharaji's only worldly possessions seemed to be the white cotton dhoti he wore on hot days and the heavy woolen plaid blanket he wrapped around himself on cold ones. He kept no particular schedule, had no organization, nor offered any fixed program of yogic poses or meditations. Like most sadhus, he was itinerant, unpredictably on the move. He mainly hung out on a tucket on the porch of whatever ashram, temple, or home he was visiting at the time.

Maharaji seemed always to be absorbed in some state of ongoing quiet rapture, and, paradoxically, at the same time was attentive to whoever was with him.[1] What struck Dan was how utterly at peace and how kind Maharaji was. Like Khunu, he took an equal interest in everyone who came—and his visitors ranged from the highest-ranking government officials to beggars.

There was something about his ineffable state of mind that Dan had never sensed in anyone before meeting Maharaji. No matter what he was doing, he seemed to remain effortlessly in a blissful, loving space, perpetually at ease. Whatever state Maharaji was in seemed not some temporary oasis in the mind, but a lasting way of being: a trait of utter wellness.

BEYOND THE PARADIGM

After two months or so making daily visits to Maharaji at the ashram, Dan and his friend Jeff (now widely known as the devotional singer Krishna Das) went traveling with another Westerner who was desperate to renew his visa after spending seven years in India living as a sadhu. That journey ended for Dan at Bodh Gaya, where he was soon to meet Khunu Lama.

Bodh Gaya, in the North Indian state Bihar, is a pilgrimage site for Buddhists the world over, and most every Buddhist country has a building in the town where its pilgrims can stay. The Burmese *vihara*, or pilgrim's rest house, had been built before the takeover by a military dictatorship that forbade Burma's citizens to travel. The vihara had lots of rooms but few pilgrims—and soon became an overnight stop for the ragged band of roaming Westerners who wandered through town.

When Dan arrived there in November 1970, he met the sole long-term American resident, Joseph Goldstein, a former Peace Corps worker in Thailand. Joseph had spent more than four years studying at the vihara with Anagarika Munindra, a meditation master. Munindra, of slight build and always clad in white, belonged to the Barua caste in Bengal, whose members had been Buddhist since the time of Gautama himself.[2]

Munindra had studied *vipassana* (the Theravadan meditation and root source of many now-popular forms of mindfulness) under Burmese masters of great repute. Munindra, who became Dan's first instructor in the method, had just invited his friend S. N. Goenka, a jovial, paunchy former businessman recently turned meditation teacher, to come to the vihara to lead a series of ten-day retreats.

Goenka had become a meditation teacher in a tradition established by Ledi Sayadaw, a Burmese monk who, as part of a cultural renaissance in the early twentieth century meant to counter British colonial influence, revolutionized meditation by making it widely available to laypeople. While meditation in that culture had for centuries been the exclusive provenance of monks and nuns, Goenka learned vipassana from U Ba Khin (U is an honorific in Burmese), at one time Burma's accountant general, who had been taught the method by a farmer, who was in turn taught by Ledi Sayadaw.

Dan took five of Goenka's ten-day courses in a row, immersing himself in this rich meditation method. He was joined by about a hundred fellow travelers. This gathering in the winter of 1970–71 was a seminal moment in the transfer of mindfulness from an esoteric practice in Asian countries to its current widespread adoption around the world. A handful of the students there, with Joseph Goldstein leading the way, later became instrumental in bringing mindfulness to the West.[3]

Starting in his college years Dan had developed a twice-daily habit of twenty-minute meditation sessions, but this immersion in ten days of continual practice brought him to new levels. Goenka's method started with simply noting the sensations of breathing in and out—not for just twenty minutes but for hours and hours a day. This cultivation of concentration then morphed into a systematic whole-body scan of whatever sensations were occurring anywhere in the body. What had been "my body, my knee" becomes a sea of shifting sensation—a radical shift in awareness.

Such transformative moments mark the boundary of mindfulness, where we observe the ordinary ebb and flow of the mind, with a further reach where we gain insight into the mind's nature. With mindfulness you would just note the stream of sensations.

The next step, insight, brings the added realization of how we claim those sensations as "mine." Insight into pain, for example, reveals how we attach a sense of "I" so it becomes "my pain" rather than being just a cacophony of sensations that change continuously from moment to moment.

This inner journey was explained in meticulous detail in mimeographed booklets of practice advice—well worn in the manner of hand-to-hand underground publications—written by Mahasi Sayadaw, Munindra's Burmese meditation teacher. The ragged pamphlets gave detailed instruction in mindfulness and stages far beyond, to further reaches of the path.

These were practical handbooks for transforming the mind with recipes for mental "hacking" that had been in continuous use for millennia.[4] When used along with one-on-one oral teachings tailored to the student, these detailed manuals could guide a meditator to mastery.

The manuals shared the premise that filling one's life with meditation and related practices produces remarkable transformations of being. And the overlap in qualities between Khunu, Maharaji, and a handful of other such beings Dan met in his travels around India seemed to affirm just such possibilities.

Spiritual literature throughout Eurasia converges in descriptions of an internal liberation from everyday worry, fixation, self-focus, ambivalence, and impulsiveness—one that manifests as freedom from concerns with the self, equanimity no matter the difficulty, a keenly alert "nowness," and loving concern for all.

In contrast, modern psychology, just about a century old, was clueless about this range of human potential. Clinical psychology, Dan's field, was fixated on looking for a specific problem like high anxiety and trying to fix that one thing. Asian psychologies had a wider lens on our lives and offered ways to enhance our positive side. Dan resolved that on his return to Harvard from India, he would make his colleagues aware of what seemed an inner upgrade far more pervasive than any dreamed of in our psychology.[5]

Just before coming to India, Dan had written an article—based on his own first flings with meditation during college and on the scant sources on the topic then available in English—that proposed the existence of such a lasting ultra-benign mode of consciousness.[6] The major states of consciousness, from the perspective of the science of the day, were waking, sleeping, and dreaming—all of which had distinctive brain wave signatures. Another kind of consciousness—more controversial and lacking any strong support in scientific evidence—was the total absorption in undistracted concentration, *samadhi* in Sanskrit, an altered state reached through meditation.

There was but one somewhat questionable scientific case study relating to samadhi that Dan could cite at the time: a report of a researcher touching a heated test tube to a yogi in samadhi, whose EEG supposedly revealed that he remained oblivious to the pain.[7]

But there was not a shred of data that spoke to any longer-lasting, benign quality of being. And so all Dan could do was hypothesize. Yet here in India, Dan met beings who just might embody that rarefied consciousness. Or so it seemed.

Buddhism, Hinduism, Jainism—all the religions that sprouted within Indian civilization—share the concept of "liberation" in one form or other. Yet psychology knows that our assumptions bias what we see. Indian culture held a strong archetype of the "liberated" person, and that lens, Dan knew, might readily foster wishful projections, a false image of perfection in the service of a pervasive and powerful belief system.

So the question remained about these rarefied qualities of being: fact or fairy tale?

THE MAKING OF A REBEL

Just as most every home in India has an altar, so do their vehicles. If it's one of the ubiquitous huge, lumbering Tata trucks, and the driver happens to be Sikh, the pictures will feature Guru Nanak, the revered founder of that religion. If a Hindu driver, there will be a deity, perhaps Hanuman, Shiva, or Durga, and usually a favorite saint or guru. That portraiture makes the driver's seat a mobile *puja* table, the sacred place in an Indian home where daily prayer occurs.

The fire-engine-red VW van that Dan drove around Cambridge after returning to Harvard from India in the fall of 1972 featured its own pantheon. Among the images Scotch-taped to the dashboard were Neem Karoli Baba, as well as other saints he had heard about: an otherworldly image of Nityananda, a radiantly smiling Ramana Maharshi, and the mustached, mildly amused visage of Meher Baba with his slogan—later popularized by singer Bobby McFerrin—"Don't worry. Be happy."

Dan had parked the van not far from the evening meeting of a course on psychophysiology he was taking to acquire the lab skills he would need for his doctoral dissertation, a study of meditation as an intervention in the body's reactions to stress. There were just a hand-ful of students seated around a seminar table in that room on the four-teenth floor of William James Hall. Richie happened to choose the chair next to Dan, and our first meeting was that night.

Talking after class, we discovered a common goal: we wanted to use our dissertation research as an opportunity to document some of the benefits that meditation brings. We were taking that psychophysi-ology seminar to learn the methods we would need.

Dan offered a ride back to the apartment Richie shared with Susan (Richie's sweetheart since college, and now his wife). Richie's reaction to the VW's dashboard puja was wide-eyed astonishment. But he was delighted to be riding with Dan: even as an undergraduate, Richie read broadly in psychology journals, including the obscure *Journal of Transpersonal Psychology*, where he had come upon Dan's article.

As Richie recalls, "It blew my mind that someone at Harvard was writing an article like that." When he was applying to grad school, he had taken this as one of several signs that he should choose Harvard. Dan, for his part, was pleased that someone had taken the article seriously.

Richie's interests in consciousness had been first aroused by the works of authors such as Aldous Huxley, British psychiatrist R. D. Laing, Martin Buber, and, later, Ram Dass, whose *Be Here Now* was published just at the start of his graduate studies.

But these interests had been driven underground during his col-lege years in the psychology department at New York University's up-town campus in the Bronx, where staunch behaviorists, followers of B. F. Skinner, dominated the psychology department.[8] Their firm as-sumption was that only observable behavior was the proper study of psychology—looking inside the mind was a questionable endeavor, a taboo waste of time. Our mental life, they held, was completely irrelevant to understanding behavior.[9]

When Richie signed up for a course in abnormal psychology, the textbook was ardently behaviorist, claiming that all psychopathology was the result of operant conditioning, where a desired behavior earns a reward, like a tasty pellet for a pigeon when it pecks the right button. That view, Richie felt, was bankrupt: it not only ignored the mind, it also ignored the brain. Richie, who could not stomach this dogma, dropped the course after the first week.

Richie's steely conviction was that psychology should study the mind—not reinforcement schedules for pigeons—and so he became a rebel. Richie's interests in what went on in the mind were, from the strict behaviorist perspective, transgressive.[10]

While by day he fought the behaviorist tide, his nights were his own to explore other interests. He volunteered to help with sleep research at Maimonides Medical Center, where he learned how to monitor brain activity with EEGs, an expertise that would serve him well throughout the rest of his career in the field.

His senior honors thesis adviser was Judith Rodin, with whom Richie conducted research on daydreaming and obesity. His hypothesis was that because daydreams take us out of the present, we become less sensitive to the body's cues of satiety, and so continue eating instead of stopping. The obesity part was because of Rodin's interest in the topic; daydreaming was Richie's way of beginning to study consciousness.[11] For Richie the study was an excuse to learn techniques to probe what was actually going on inside the mind, using physiological and behavioral measures.

Richie monitored people's heart rate and sweating while they let their mind wander or did mental tasks. This was his first use of physiological measures to infer mental processes, a radical method at the time.[12]

This methodological sleight of hand, tacking an element of consciousness studies on to an otherwise respectable, mainstream research study, was to be a hallmark of Richie's research for the next decade or so, when his interest in meditation found little to no support in the ethos of the time.

Designing a dissertation that didn't depend on the meditation piece in itself but could be a stand-alone study on just the nonmeditators turned out to be a smart move for Richie. He secured his first academic position at the Purchase campus of the State University of New York, where he kept his interest in meditation to himself while doing seminal work in the emerging field of affective neuroscience—how emotions operate in the brain.

Dan, however, could find no teaching post at any university that reflected his own interests in consciousness, and gladly accepted a job in journalism—a career path that eventually led to his becoming a science writer at the *New York Times*. While there he harvested Richie's research on emotions and the brain (among other scientists' work) in writing *Emotional Intelligence*.[13]

Of the more than eight hundred articles Dan wrote at the *Times*, just a meager handful had anything to do with meditation—even as we both continued to attend meditation retreats on our own time. We shelved the notion publicly for a decade or two, while privately pursuing the evidence that intense and prolonged meditation can alter the core of a person's very being. We were both flying under the radar.

ALTERED STATES

William James Hall looms over Cambridge as an architectural mistake, a fifteen-story modernist white slab glaringly out of place amid the surrounding Victorian homes and the low-lying brick-and-stone buildings of the Harvard campus. At the beginning of the twentieth century, William James became Harvard's first professor of psychology, a field he had a major hand in inventing as he transitioned from the theoretical universe of philosophy to a more empirical and pragmatic view of the mind. James's former home still stands in the adjacent neighborhood.

Despite this history, as graduate students in the department housed in William James Hall, we were never assigned a single page of James to read—he had long before fallen out of fashion. Still, James became an inspiration to us, largely because he engaged the very topic that our professors ignored and that fascinated us: consciousness.

Back in James's day, toward the end of the nineteenth and start of the twentieth centuries, there was a fad among Boston's cognoscenti to imbibe nitrous oxide (or "laughing gas," as the compound came to be called when dentists routinely deployed it). James's transcendent moments with the help of nitrous oxide led him to what he called an "unshakable conviction" that "our normal waking consciousness . . . is but one special type of consciousness, whilst all about it, parted from it by the filmiest of screens, there lie potential forms of consciousness entirely different."[14]

After pointing out the existence of altered states of consciousness (though not by that name), James adds, "We may go through life without suspecting their existence; but apply the requisite stimulus, and at a touch they are there in all their completeness."

Dan's article had begun with this very passage from William James's *The Varieties of Religious Experience*, a call to study altered states of consciousness. These states, as James saw, are discontinuous with ordinary consciousness. And, he observed, "No account of the universe in its totality can be final which leaves these other forms of consciousness quite disregarded." The very existence of these states "means they forbid a premature closing of our accounts with reality."

Psychology's topography of the mind foreclosed such accounts. Transcendental experiences were not to be found anywhere in that terrain; if mentioned at all, they were relegated to the less desirable realms. From the early days of psychology, beginning with Freud himself, altered states were dismissed as symptoms of one or another form of psychopathology. For instance, when French poet and Nobel laureate Romain Rolland became a disciple of the Indian saint Sri Ramakrishna around the beginning of the twentieth century, he wrote to Freud describing the mystical state he experienced—and Freud diagnosed it as regression to infancy.[15]

By the 1960s, psychologists routinely dismissed drug-triggered altered states as artificially induced psychosis (the original term for psychedelics was "psychotomimetic" drugs—psychosis mimics). As we found, similar attitudes applied to meditation—this suspicious new route to altering the mind—at least among our faculty advisers.

Still, in 1972 the Cambridge zeitgeist included a fervent interest in consciousness as Richie entered Harvard and Dan returned from his sojourn in Asia (the first of two) to begin his doctoral dissertation. Charles Tart's bestseller of the day, *Altered States of Consciousness*, collected articles on biofeedback, drugs, self-hypnosis, yoga, meditation, and other such avenues to James's "other states," capturing the ethos of the day.[16] In brain science, excitement revolved around the recent

discovery of neurotransmitters, the chemicals that send messages between neurons, like the mood regulator serotonin—magic molecules
that could pitch us into ecstasy or despair.[17]

The lab work on neurotransmitters filtered into the general culture as a scientific pretext for attaining altered states through drugs
like LSD. These were the days of the psychedelic revolution, which
had had its roots in the very department at Harvard we were in, which
perhaps helps explain why the remaining stalwarts took a dim view of
any interest in the mind that smacked of altered states.

AN INNER JOURNEY

Dalhousie nestles in the lower reaches of the Dhauladhar range, a
branch of the Himalayas that stretches into India's Punjab and Hi-
machal Pradesh states. Established in the mid-nineteenth century as a
"hill station" where the bureaucrats of the British Raj could escape the
summer heat of the Indo-Gangetic Plain, Dalhousie was chosen for
its gorgeous setting. With its picturesque bungalows left over from
colonial days, this hill station has long been a tourist attraction.

But it wasn't the setting that brought Richie and Susan to Dalhousie that summer of 1973. They had come for a ten-day retreat—their
first deep dive—with S. N. Goenka, the same teacher Dan had done
successive retreats with in Bodh Gaya a few years before while on his
first sojourn in India for his predoctoral traveling fellowship. Richie and
Susan had just visited Dan in Kandy, Sri Lanka, where he was living on
a postdoctoral fellowship during this second trip to Asia.[18]

Dan encouraged the couple to take a course with Goenka as a
doorway into intensive meditation. The course was a bit disorienting

from the start. For one, Richie slept in a large tent for the men, Susan in one for the women. And the imposition of "noble silence" from day one meant that Richie never really knew who else shared that tent—his vague impression was that they were mostly Europeans.

In the meditation hall Richie found the floor scattered with round zafus, Zen-style cushions, to sit on. The zafu would be Richie's perch through the twelve or so hours of sitting in meditation the daily schedule called for.

Settling onto his zafu in his usual half lotus, Richie noticed a twinge of pain in his right knee, which had always been the weak one. As the hours of sitting progressed day by day, that twinge morphed into a low howl of discomfort, and spread not just to the other knee but to his lower back as well—common hurt zones for Western bodies unaccustomed to sitting still for hours supported by nothing but a pillow on the floor.

Richie's mental task for the whole day was to tune in to the sensations of breathing at his nostrils. The most vivid sense impression wasn't his breath—it was the continual intense physical pain in his knees and back. By the end of the first day, he was thinking, I can't believe I have nine more days of this.

But on the third day came a major shift with Goenka's instruction to "sweep" with a careful, observing attention head to toe, toe to head, through all the many and varied sensations in his body. Though Richie found his focus returning again and again to the throbbing pain in that knee, he also started to glimpse a sense of equanimity and well-being.

Soon Richie found himself entering a state of total absorption that, toward the end of the retreat, allowed him to sit for up to four hours at a go. At lights-out time he'd go to the empty meditation hall and meditate on his body's sensations steadily, sometimes until 1:00 or 2:00 a.m.

The retreat was a high for Richie. He came away with a deep conviction that there were methods that could transform our minds to produce a profound well-being. We did not have to be controlled by the mind, with its random associations, sudden fears and angers, and all the rest—we could take back the helm.

For days after the retreat ended, Richie still felt he was on a high. Richie's mind kept soaring while he and Susan stayed on in Dalhousie. The high rode with him on the bus down the mountains via roads wending through fields and villages with mud-walled, thatch-roofed houses, on to the busier cities of the plains, and finally through the throbbing, packed roads of Delhi.

There Richie felt that high begin to wane as he and Susan spent a few days in the bare-bones guesthouse they could afford on their grad student budget, venturing out to Delhi's cacophonous and crowded streets to have a tailor make some clothes and buy souvenirs.

Perhaps the biggest force in the decline of that meditation state was the traveler's stomach they both had come down with. That malady plagued them through a change of planes in Frankfurt on the cheap flight from Delhi to Kennedy Airport. After a full day spent in travel they landed in New York, where they were greeted by both sets of parents, eager to see them after this summer away in Asia.

As Susan and Richie exited Customs—sick, tired, and dressed in the Indian style of the day—their families greeted them with looks of horrified shock. Instead of enveloping them in love, they yelled in alarm, "What have you done to yourselves? You look terrible!"

By the time they all arrived at the upstate New York country house of Susan's family, the half-life of that high had reached the bottom of its slope, and Richie felt as terrible as he'd looked walking off the plane.

Richie tried to revive the state he had reached at the Dalhousie course, but it had vanished. It reminded him of a psychedelic trip in that way: he had vivid memories of the retreat, but they were not embodied, not a lasting transformation. They were just memories.

That sobering experience fed into what was to become a burning scientific question: How long do state effects—like Richie's meditative highs—last? At what point can they be considered enduring traits? What allows such a transformation of being to become embodied in a lasting way instead of fading into the mists of memory?

And just where in the mind's terrain had Richie been?

A MEDITATOR'S GUIDEBOOK

The bearings for Richie's inner whereabouts were more than likely to be detailed somewhere in a thick volume that Munindra had encouraged Dan to study during his first sojourn in India a few years before: the *Visuddhimagga*. This fifth-century text, which means *Path to Purification* in Pali (the language of Buddhism's earliest canon), was the ancient source for those mimeographed manuals Dan had pored over in Bodh Gaya.

Though centuries old, the *Visuddhimagga* remained the definitive guidebook for meditators in places like Burma and Thailand, that follow the Theravada tradition, and through modern interpretations still offers the fundamental template for insight meditation, the root of what's popularly known as "mindfulness."

This meditator's manual on how to traverse the mind's most

subtle regions offered a careful phenomenology of meditative states and their progression all the way to nirvana (*nibbana*, in Pali). The highways to the jackpot of utter peace, the manual revealed, were a keenly concentrated mind on the one hand, merging with a sharply mindful awareness on the other.

The experiential landmarks along the way to meditative attainments were spelled out matter-of-factly. For instance, the path of concentration begins with a mere focus on the breath (or any of more than forty other suggested points of focus, such as a patch of color—anything to focus the mind). For beginners this means a wobbly dance between full focus and a wandering mind.

At first the flow of thoughts rushes like a waterfall, which sometimes discourages beginners, who feel their mind is out of control. Actually, the sense of a torrent of thoughts seems to be due to paying close attention to our natural state, which Asian cultures dub "monkey mind," for its wildly frenetic randomness.

As our concentration strengthens, wandering thoughts subside rather than pulling us down some back alley of the mind. The stream of thought flows more slowly, like a river—and finally rests in the stillness of a lake, as an ancient metaphor for settling the mind in meditation practice tells us.

Sustained focus, the manual notes, brings the first major sign of progress, "access concentration," where attention stays fixed on the chosen target without wandering off. With this level of concentration come feelings of delight and calm, and, sometimes, sensory phenomena like flashes of light or a sense of bodily lightness.

"Access" implies being on the brink of total concentration, the full absorption called *jhana* (akin to samadhi in Sanskrit), where any and all distracting thoughts totally cease. In jhana the mind fills with strong rapture, bliss, and an unbroken one-pointed focus on the meditation target.

The *Visuddhimagga* lists seven more levels of jhana, with progress marked by successively subtle feelings of bliss and rapture, and stronger equanimity, along with an increasingly firm and effortless focus. In the last four levels, even bliss, a relatively gross sensation, falls away, leaving only unshakable focus and equanimity. The highest reach of this ever more refined awareness has such subtlety it is called the jhana of "neither perception nor nonperception."

In the time of Gautama Buddha, full concentrated absorption in samadhi was heralded as the highway to liberation for yogis. Legend has it that the Buddha practiced this approach with a group of wandering ascetics, but he abandoned that avenue and discovered an innovative variety of meditation: looking deeply into the mechanics of consciousness itself.

Jhana alone, the Buddha is said to have declared, was not the path to a liberated mind. Though strong concentration can be an enormous aid along the way, the Buddha's path veers into a different kind of inner focus: the path of insight.

Here, awareness stays open to whatever arises in the mind rather than to one thing only—to the exclusion of all else—as in total concentration. The ability to maintain this mindfulness, an alert but nonreactive stance in attention, varies with our powers of one-pointedness.

With mindfulness, the meditator simply notes without reactivity whatever comes into mind, such as thoughts or sensory impressions like sounds—and lets them go. The operative word here is *go*. If we think much of anything about what just arose, or let it trigger any reactivity at all, we have lost our mindful stance—unless that reaction or thought in turn becomes the object of mindfulness.

The *Visuddhimagga* describes the way in which carefully sustained mindfulness—"the clear and single-minded awareness of what actually happens" in our experience during successive moments—refines into a more nuanced insight practice that can lead us through a succession of stages toward that final epiphany, nirvana/*nibbana*.[19]

This shift to insight meditation occurs in the relationship of our awareness to our thoughts. Ordinarily our thoughts compel us: our loathing or self-loathing generates one set of feelings and actions; our romantic fantasies quite another. But with strong mindfulness we can experience a deep sense in which self-loathing and romantic thoughts are the same: like all other thoughts, these are passing moments of mind. We don't have to be chased through the day by our thoughts—they are a continuous series of short features, previews, and outtakes in a theater of the mind.

Once we glimpse our mind as a set of processes, rather than getting swept away by the seductions of our thoughts, we enter the path of insight. There we progress through shifting again and again our relationship to that inner show—each time yielding yet more insights into the nature of consciousness itself.

Just as mud settling in a pond lets us see into the water, so the subsiding of our stream of thought lets us observe our mental machinery with greater clarity. Along the way, for instance, the meditator sees a bewilderingly rapid parade of moments of perception that race through the mind, ordinarily hidden from awareness somewhere behind a scrim.

Richie's meditation high most certainly could be spotted somewhere in these benchmarks of progress. But that high had disappeared into the mists of memory. *Sic transeunt* altered states.

In India they tell of a yogi who spent years and years alone in a cave, achieving rarefied states of samadhi. One day, satisfied that he had reached the end of his inner journey, the yogi came down from his mountain perch into a village.

That day the bazaar was crowded. As he made his way through the crowd, the yogi was caught up in a rush to make way for a local lord riding through on an elephant. A young boy standing in front of the yogi stepped back suddenly in fright—stomping right on the yogi's bare foot.

The yogi, angered and in pain, raised his walking staff to strike the youngster. But suddenly seeing what he was about to do—and the anger that propelled his arm—the yogi turned around and went right back up to his cave for more practice.

The tale speaks to the difference between meditation highs and enduring change. Beyond transitory states like samadhi (or their equivalent, the absorptive jhanas), there can be lasting changes in our very being. The *Vissudhimagga* holds this transformation to be the true fruit of reaching the highest levels of the path of insight. For example, as the text says, strong negative feelings like greed and selfishness, anger and ill will, fade away. In their place comes the predominance of positive qualities like equanimity, kindness, compassion, and joy.

That list resonates with similar claims from other meditative traditions. Whether these traits are due to some specific transformative experiences that accrue in attaining those levels, or from the sheer hours of practice along the way, we can't say. But Richie's delicious meditation-induced high—possibly somewhere in the vicinity of access concentration, if not first jhana—was not sufficient to bring on these trait changes.

The Buddha's discovery—reaching enlightenment via the path of insight—was a challenge to the yogic traditions of his day, which followed the path of concentration to various levels of samadhi, the bliss-filled state of utter absorption. In those days, insight versus concentration was a burning issue in a politics of consciousness that revolved around the best path to those altered traits.

Fast-forward to another politics of consciousness in the 1960s, during the heady days of the psychedelic fad. The sudden revelations of drug-induced altered states led to assumptions like, as one acidhead put it, "With LSD we experienced what it took Tibetan monks 20 years to obtain, yet we got there in 20 minutes."[20]

Dead wrong. The trouble with drug-induced states is that after the chemical clears your body, you remain the same person as always. And, as Richie discovered, the same fading away happens with highs in meditation.

3

The After Is the Before for the Next During

Dan's second stay in Asia was in 1973, this time on a Social Science Research Council postdoc, ostensibly a venture in "ethnopsychology," to study Asian systems for analyzing the mind and its possibilities. It started with six months in Kandy, a town in the hills of Sri Lanka where Dan consulted every few days with Nyanaponika Thera, a German-born Theravadan monk whose scholarship centered on the theory and practice of meditation. (Dan then continued on for several months in Dharamsala, India, where he studied at the Library of Tibetan Works and Archives.)

Nyanaponika's writings focused on the *Abhidhamma*, a model of mind that laid out a map and methods for the transformation of consciousness in the direction of altered traits. While the *Visuddhimagga* and the meditation manuals Dan had read were operator's instructions for the mind, the *Abhidhamma* was a guiding theory for such manuals. This psychological system came with a detailed explanation of the mind's key elements and how to traverse this inner landscape to make lasting changes in our core being.

Certain sections were compelling in their relevance to psychology, particularly the dynamic outlined between "healthy" and "unhealthy" states of mind.[1] All too often our mental states fluctuate in a range that highlights desires, self-centeredness, sluggishness, agitation, and the like. These are among the unhealthy states on this map of mind.

Healthy states, in contrast, include even-mindedness, composure, ongoing mindfulness, and realistic confidence. Intriguingly, a subset of healthy states applies to both mind and body: buoyancy, flexibility, adaptability, and pliancy.

The healthy states inhibit the unhealthy ones, and vice versa. The mark of progress along this path is whether our reactions in daily life signal a shift toward healthy states. The goal is to establish the healthy states as predominant, lasting traits.

While immersed in deep concentration, a meditator's unhealthy states are suppressed—but, as with that yogi in the bazaar, can emerge as strong as ever when the concentrative state subsides. In contrast, according to this ancient Buddhist psychology, attaining deepening levels of insight practice leads to a radical transformation, ultimately freeing the meditator's mind of the unhealthy mix. A highly advanced practitioner effortlessly stabilizes on the healthy side, embodying confidence, buoyancy, and the like.

Dan saw this Asian psychology as a working model of the mind, time-tested over the course of centuries, a theory of how mental training could lead to highly positive altered traits. That theory had guided meditation practice for more than two millennia—it was an electrifying proof of concept.

In the summer of 1973, Richie and Susan came to Kandy for a six-week visit before heading to India for that thrilling and sobering retreat with Goenka. Once together in Kandy, Richie and Dan trekked through the jungle to consult with Nyanaponika at his remote hermitage about this model of mental well-being.[2]

Later that year, after Dan returned from this second sojourn in Asia as a Social Science Research Fellow, he was hired at Harvard as a visiting lecturer. In the fall semester of 1974 he offered a course, The Psychology of Consciousness, which fit well the ethos of those days—at least among students, many of whom were doing their own extracurricular research with psychedelics, yoga, and even a bit of meditation.

Once the psychology of consciousness course was announced, hundreds of Harvard undergrads gravitated to this survey of meditation and its altered states, the Buddhist psychological system, and what little was then known about the dynamics of attention—all among the topics covered. The enrollment was so large that the class was moved into the largest classroom venue at Harvard, the 1,000-seat Sanders Theatre.[3] Richie, then in his third year of graduate school, was a teaching assistant in the course.[4]

Most of the topics in The Psychology of Consciousness—and the course title itself—were far outside the conventional map of psychology in those days. No surprise, Dan was not asked to stay on by the department after that semester finished. But by then we had done some writing and research together, and Richie was excited by the realization that this was what his own research path would be and was eager to get going.

Starting while we were in Sri Lanka and continuing during Dan's semester teaching that course on the psychology of consciousness, we worked on the first draft of our article, making the case to our colleagues in psychology for altered traits. While Dan had, of necessity, based his first article on thin claims, scant research, and much guesswork, now we had a template for the path to altered traits, an algorithm for inner transformation. We wrestled with how to connect this map with the sparse data science had by then yielded.

Back in Cambridge we mulled all this over in long conversations, often in Harvard Square. As vegetarians at the time, we settled on caramel sundaes at Bailey's ice cream parlor on Brattle Street. There we worked on what would become a journal article piecing together the little relevant data we could find to support our first statement of extremely positive altered traits.

We called it "The Role of Attention in Meditation and Hypnosis: A Psychobiological Perspective on Transformations of Consciousness." The operative phrase here is *transformations of consciousness*, our term then for altered traits, which we saw as a "psychobiological" (today we'd say "neural") shift. We contended that hypnosis, unlike meditation, produced primarily state effects, and not trait effects as with meditation.

In those times the fascination was not with traits but rather altered states, whether from psychedelics or meditation. But, as we put it in talking at Bailey's, "after the high goes, you're still the same schmuck you were before." We articulated the idea more formally in the subsequent journal article.

We were speaking to a basic confusion, still too common, about how meditation can change us. Some people fixate on the remarkable states attained during a meditation session—particularly during long retreats—and give little notice to how, or even if, those states translate into a lasting change for the better in their qualities of being after they've gone home. Valuing just the heights misses the true point of practice: to transform ourselves in lasting ways day to day.

More recently, this point was driven home to us when we had the chance to tell the Dalai Lama about the meditative states and their brain patterns that a longtime practitioner displayed in Richie's lab. As this expert engaged in specific kinds of meditation—for instance, concentration or visualization—the brain imaging data revealed a distinct neural profile for each meditative altered state.

"It's very good," the Dalai Lama commented, "he managed to show some signs of yogic ability"—by which he meant the intensive meditation over months or years practiced by yogis in Himalayan caves, as opposed to the garden variety of yoga for fitness so popular these days.[5]

But then he added, "The true mark of a meditator is that he has disciplined his mind by freeing it from negative emotions."

That rule of thumb has stayed constant since before the time of the *Visuddhimagga*: It's not the highs along the way that matter. It's who you become.

Puzzling over how to reconcile the meditation map with what we had experienced ourselves, and then with the admittedly scant scientific evidence, we articulated a hypothesis: *The after is the before for the next during.*

To unpack this idea, *after* refers to enduring changes from meditation that last long beyond the practice session itself. *Before* means the condition we are in at baseline, before we start meditating. *During* is what happens as we meditate, temporary changes in our state that pass when we stop meditating.

In other words, repeated practice of meditation results in lasting traits—the *after*.

We were intrigued by the possibility of some biological pathway where repeated practice led to a steady embodiment of highly positive traits like kindness, patience, presence, and ease under any circumstances. Meditation, we argued, was a tool to foster precisely such beneficial fixtures of being.

We published our article in one of maybe two or three academic publications interested in such exotic topics as meditation back in the 1970s.[6] This was a first glimmer of our thinking on altered traits, albeit with a flimsy science base. The maxim "probability is not proof" applied, in a sense: what we had was a possibility, but little to pin a probability on, and zero proof.

When we first wrote about this, no scientific study had been conducted that would provide the kind of evidence we needed. Only long decades after we published the article would Richie find that for highly adept meditators, their "before" state was, indeed, very different from that of people who had never meditated, or done very little meditating—it was an indicator of an altered trait (as we'll see in chapter twelve, "Hidden Treasure").

No one in psychology in those days had talked about altered *traits*. Plus, our raw material was highly unusual for psychologists: ancient meditation manuals, then hard to come by outside Asia, along with our own experiences in intensive meditation retreats, and chance meetings with highly adept practitioners. We were, to say the least, outliers in psychology—or oddballs, as we no doubt were perceived by some of our Harvard colleagues.

Our vision of altered traits made a leap far beyond the psychological science of our day. Risky business.

THE SCIENCE CATCHES UP

When an imaginative researcher concocts a novel idea, it starts a chain of events much like natural variation in evolution: as sound empirical tests weigh new ideas, they eliminate bad hypotheses and spread good ones.[7]

For this to happen, science needs to balance skeptics with speculators—people who cast wide nets, think imaginatively, and consider "what if." The web of knowledge grows by testing original ideas brought to it by speculators like ourselves. If only skeptics pursued science, little innovation would occur.

Economist Joseph Schumpeter has become known these days for the concept of "creative destruction," where the new disrupts the old in a market. Our early hunches about altered traits fit what Schumpeter called "vision": an intuitive act that supplies direction and energy for analytic efforts. A vision lets you see things in a new light, as he says, one "not to be found in the facts, methods, and results of the preexisting state of the science."[8]

Sure, we had a vision in this sense—but we had paltry methods or data available for exploring this positive range of altered traits, and no idea of the brain mechanism that would allow such a profound shift. We were determined to make the argument, but were years too soon for the crucial scientific piece in this puzzle.

Our dissertation data were feebly—*very* feebly—supportive of the idea that the more you practice how to generate a meditative state, the more that practice shows lasting influences beyond the session itself.

Still, as brain science has evolved over the decades, we saw mounting rationales for our ideas.

Richie attended his first meeting of the Society for Neuroscience in 1975 in New York City, along with about 2,500 other scientists, all exhilarated that they were seeing the birth of a new field (and none dreaming that these days those meetings would draw more than 30,000 neuroscientists).[9] In the mid-1980s one of the early presidents of the society, Bruce McEwen of Rockefeller University, gave us scientific ammunition.

McEwen put a dominant tree shrew in the same cage for twenty-eight days with one lower in the pecking order—the rodent version of being trapped at work with a nightmare boss 24/7 for a month. The big shock from McEwen's study was that in the brain of the dominated rodent, dendrites shrank in the hippocampus, a node crucial for memory. These branching projections of the body's cells allow them to reach out to and act on other cells; shrinking dendrites mean faulty memory.

McEwen's results ripped through the brain and behavioral sciences like a small tsunami, opening minds to the possibility that a given experience could leave an imprint on the brain. McEwen was zeroing in on a holy grail for psychology: how stressful events produce lingering neural scars. That an experience of any kind could leave its mark on the brain had, until then, been unthinkable.

To be sure, stress was par for the course for a laboratory rat—McEwen just upped the intensity. The standard setup for lab rat living quarters was the rodent equivalent of solitary confinement: weeks or months on end in a small wire cage and, if the rat was lucky, a running wheel for exercise.

Contrast that life in perpetual boredom and social isolation to something like a rodent health resort, with lots of toys, things to climb on, colorful walls, playmates, and interesting spaces to explore. That's the stimulating habitat Marion Diamond at the University of California at Berkeley built for her lab rats. Working about the same time as McEwen, Diamond found the rats' brains benefited, with thicker dendritic branches connecting neurons and growth in brain areas, such as the prefrontal cortex, that are crucial in attention and self-regulation.[10]

While McEwen's work showed how adverse events can shrink parts of the brain, Diamond's emphasized the positive in her studies. Yet her work was largely met with a shrug in neuroscience, perhaps because it posed a direct challenge to pervasive beliefs in the field. The conventional wisdom then was that at birth we host in our skull a maximum number of neurons, and then inexorably lose them in a steady die-off over the course of life. Experience, supposedly, had nothing to do with this.

But McEwen and Diamond led us to wonder, If these brain changes for worse and for better could occur with rats, might the right experience change the human brain toward beneficial altered traits? Could meditation be just such a helpful inner workout?

The glimpse of this possibility was exhilarating. We sensed something truly revolutionary was in the offing, but it took a couple more decades before the evidence began to catch up with our hunch.

THE BIG LEAP

The year was 1992, and Richie was nervous when the sociology department at the University of Wisconsin asked him to deliver a major departmental colloquium. He knew he was walking into the center of an intellectual cyclone, a battle over "nature" and "nurture" that had raged for years in the social sciences. The nurture camp believed that our behavior was shaped by our experiences; the "nature" camp saw our genes as determining our behavior.

The battle had a long, ugly history—racists in the nineteenth and early twentieth centuries twisted the genetics of their day as "scientific" grounds for bias against blacks, Native Americans, Jews, the Irish, and a long list of other targets of bigotry. The racists attributed any and all lags in educational and economic attainments of the target group to their genetic destiny, ignoring vast imbalances in opportunity. The resulting backlash in the social sciences had made many in that sociology department deeply skeptical of any biological explanation.

But Richie felt that sociologists committed a scientific fallacy in immediately assuming that biological causes necessarily reduced group differences to genetics—and so were seen as unchangeable. In Richie's view, these sociologists were carried away by an ideological stance.

For the first time in public he proposed the concept of "neuroplasticity" as a way to resolve this battle between nature and nurture. Neuroplasticity, he explained, shows that repeated experience can change the brain, shaping it. We don't have to choose between nature *or* nurture. They interact, each molding the other.

The concept neatly reconciled what had been hostile points of view. But Richie was reaching beyond the science of the day; the data on human neuroplasticity were still hazy.

That changed just a few years later with a cascade of scientific findings—for instance, those showing that mastering a musical instrument enlarged the relevant brain centers.[11] Violinists, whose left hands continuously fingered the strings while they played, had enlarged areas of the brain that manage that finger work. The longer they had played, the greater the size.[12]

NATURE'S EXPERIMENT

Try this. Look straight ahead and hold up a finger with your arm outstretched. Still looking straight ahead, slowly shift that finger until it is about two feet to the right of your nose. When you move your finger far to the right, but stay focused straight ahead, it lands in your peripheral vision, the outer edge of what your visual system takes in.[13]

Most people lose sight of their finger as it moves to the far right or left of their nose. But one group does not: people who are deaf.

While this unusual visual advantage in the deaf has long been known, the brain basis has only recently been shown. And the mechanism is, again, neuroplasticity.

Brain studies like this take advantage of so-called "experiments of nature," naturally occurring situations such as congenital deafness. Helen Neville, a neuroscientist at the University of Oregon with a passionate interest in brain plasticity, seized the opportunity to use an MRI brain scanner to test both deaf and hearing people with a visual simulation that mimicked what a deaf person sees when reading sign language.

Signs are expansive gestures. When a deaf person is reading the signing of another, she typically looks at the face of the person who is signing—not directly at how the hands move as they sign. Some of those expansive gestures move in the periphery of the visual field, and thus naturally exercise the brain's ability to perceive within this outer rim of vision. Plasticity lets these circuits take on a visual task as the deaf person learns sign language: reading what's going on at the very edge of vision.

The chunk of neural real estate that usually operates as the primary auditory cortex (known as Heschl's gyrus) receives no sensory inputs in deaf people. The brains of deaf people, Neville discovered, had morphed so that what is ordinarily a part of the auditory system was now working with the visual circuitry.[14]

Such findings illustrate how radically the brain can rewire itself in response to repeated experiences.[15] The findings in musicians and in the deaf—and a slew of others—offered a proof we had been waiting for. Neuroplasticity provides an evidence-based framework and a language that makes sense in terms of current scientific thinking.[16] It was the scientific platform we had long needed, a way of thinking about how intentional training of the mind, like meditation, might shape the brain.

THE ALTERED TRAIT SPECTRUM

Altered traits map along a spectrum starting at the negative end, with post-traumatic stress disorder (PTSD) as a case in point. The amygdala acts as the neural radar for threat. Overwhelming trauma resets to a hair trigger the amygdala's threshold for hijacking the rest of the brain to respond to what it perceives as an emergency.[17] In people with PTSD, any cue that reminds them of the traumatic experience—and that for someone else would not be particularly noticeable—sets off a cascade of neural overreactions that create the flashbacks, sleeplessness, irritability, and hypervigilant anxiety of that disorder.

Moving along the trait spectrum toward the positive range, there are the beneficial neural impacts of being a secure child, whose brain gets molded by empathic, concerned, and nurturing parenting. This childhood brain shaping builds in adulthood, for example, into being able to calm down well when upset.[18]

Our interest in altered traits looks beyond the merely healthy spectrum to an even more beneficial range, wholesome traits of being. These *extremely positive altered traits*, like equanimity and compassion, are a goal of mind training in contemplative traditions. We use the term *altered trait* as shorthand for this highly positive range.[19]

Neuroplasticity offers a scientific basis for how repeated training could create those lasting qualities of being we had encountered in a handful of exceptional yogis, swamis, monks, and lamas. Their altered traits fit ancient descriptions of lasting transformation at the higher levels.

A mind free from disturbance has value in lessening human suffering, a goal shared by science and meditative paths alike. But apart from lofty heights of being, there's a more practical potential within reach of every one of us: a life best described as flourishing.

FLOURISHING

As Alexander the Great was leading his armies through what is now Kashmir, legend has it he met a group of ascetic yogis in Taxila, then a thriving city on a branch of the Silk Road leading to the plains of India.

The yogis responded to the appearance of Alexander's fierce soldiers with indifference, saying that he, like them, could actually possess only the ground on which he stood—and that he, like them, would die one day.

The Greek-derived word for these yogis is gymnosophists, literally "naked philosophers" (even today some groups of Indian yogis roam naked, coating themselves in ashes). Alexander, impressed by their equanimity, deemed them to be "free men," and even convinced one yogi, Kalyana, to accompany him on his journey of conquest. No doubt the yogi's lifestyle and outlook resonated with Alexander's own schooling. Alexander had been tutored by the Greek philosopher Aristotle. Renowned for his lifelong love of learning, Alexander would have recognized the yogis as exemplars of another source of wisdom.

The Greek schools of philosophy espoused an ideal of personal transformation that remarkably echoes those of Asia, as Alexander may have found in his exchanges with Kalyana. The Greeks and their heirs the Romans, of course, laid the foundation for Western thought down to the present day.

Aristotle posited the goal of life as a virtue-based *eudaimonia*—a quality of flourishing—a view that continues under many guises in modern thought. Virtues, Aristotle said, are attained in part by finding the "right mean" between extremes; courage lies between impulsive risk-taking and cowardice, a tempered moderation between self-indulgence and ascetic denial.

And, he added, we are not by nature virtuous but all have the potential to become so through the right effort. That effort includes what today we would call self-monitoring, the ongoing practice of noting our thoughts and acts.

Other Greco-Roman philosophic schools used similar practices in their own paths toward flourishing. For the Stoics, one key was seeing that our feelings about life's events, not those events themselves, determine our happiness; we find equanimity by distinguishing what we can control in life from what we cannot. Today that creed finds an echo in the popularized Twelve Step version of theologian Reinhold Niebuhr's prayer:

> God, grant me the serenity to accept the things I cannot change,
> Courage to change the things I can,
> And wisdom to know the difference.

The classical way to the "wisdom to know the difference" lay in mental training. These Greek schools saw philosophy as an applied art and taught contemplative exercises and self-discipline as paths to flourishing. Like their peers to the East, the Greeks saw that we can cultivate qualities of mind that foster well-being.

The Greek practices for developing virtues were to some extent taught openly, while others were apparently given only to initiates like Alexander, who noted that the philosopher's texts were more fully understood in the context of these secretive teachings.

In the Greco-Roman tradition, qualities such as integrity, kindness, patience, and humility were considered keys to enduring well-being. These Western thinkers and Asian spiritual traditions alike saw the value in cultivating a virtuous life via a roughly similar transformation of being. In Buddhism, for example, the ideal of inner flourishing gets put in terms of *bodhi* (in Pali and Sanskrit), a path of self-actualization that nourishes "the very best within oneself."[20]

ARISTOTLE'S DESCENDANTS

Today's psychology uses the term *well-being* for a version of the Aristotelian meme *flourishing*. University of Wisconsin psychologist (and

Richie's colleague there) Carol Ryff, drawing on Aristotle among many other thinkers, posits a model of well-being with six arms:

- *Self-acceptance*, being positive about yourself, acknowledging both your best and not-so-good qualities, and feeling fine about being just as you are. This takes a nonjudgmental self-awareness.

- *Personal growth*, the sense you continue to change and develop toward your full potential—getting better as time goes on—adopting new ways of seeing or being and making the most of your talents. "Each of you is perfect the way you are," Zen master Suzuki Roshi told his students, adding, "and you can use a little improvement"—neatly reconciling acceptance with growth.

- *Autonomy*, independence in thought and deed, freedom from social pressure, and using your own standards to measure yourself. This, by the way, applies most strongly in individualistic cultures like Australia and the United States, as compared with cultures like Japan, where harmony with one's group looms larger.

- *Mastery*, feeling competent to handle life's complexities, seizing opportunities as they come your way, and creating situations that suit your needs and values.

- *Satisfying relationships*, with warmth, empathy, and trust, along with mutual concern for each other and a healthy give-and-take.

- *Life purpose*, goals and beliefs that give you a sense of meaning and direction. Some philosophers argue that true happiness comes as a by-product of meaning and purpose in life.

Ryff sees these qualities as a modern version of *eudaimonia*— Aristotle's "highest of all human good," the realization of your unique potential.[21] As we will see in the chapters that follow, different varieties of meditation seem to cultivate one or more of these capacities. More immediately, several studies have looked at how meditation boosted people's ratings on Ryff's own measure of well-being.

Fewer than half of Americans, according to the Centers for Disease Control and Prevention, report feeling a strong sense of purpose in life beyond their jobs and family obligations.[22] That particular aspect of well-being may have significant implications: Viktor Frankl has written about how a sense of meaning and purpose allowed him and select others to survive years in a Nazi concentration camp while thousands were dying around them.[23] For Frankl, continuing his work as a psychotherapist with other prisoners in the camp lent purpose to his life; for another man there, it was having a child who was on the outside; yet another found purpose in the book he wanted to write.

Frankl's sentiment resonates with a finding that after a three-month meditation retreat (about 540 hours total), those practitioners who had strengthened a sense of purpose in life during that time also showed a simultaneous increase in the activity of telomerase in their immune cells, even five months later.[24] This enzyme protects the length of telomeres, the caps at the ends of DNA strands that reflect how long a cell will live.

It's as though the body's cells were saying, stick around—you've got important work to do. On the other hand, as these researchers note, this finding needs to be replicated in well-designed studies before we can be more sure.

Also of interest: eight weeks of a variety of mindfulness seemed to enlarge a region in the brain stem that correlated with a boost in well-being on Ryff's test.[25] But the study was quite small—just fourteen people—and so, needs to be redone with a larger group before we can draw more than tentative conclusions.

Similarly, in a separate study, people practicing a popular form of mindfulness reported higher levels of well-being and other such benefits up to a year later.[26] The more everyday mindfulness, the greater the subjective boost in well-being. Again, the numbers in this study were small, and a brain measure—which, as we've said, is far less susceptible to psychological skew than self-evaluations—would be even more convincing.

So, while we find the conclusion that meditation enhances well-being an appealing idea, especially as meditators ourselves, our science side remains skeptical.

Studies such as these are often cited as "proving" the merits of meditation, particularly these days, when mindfulness has become the flavor du jour. But meditation research varies enormously when it comes to scientific soundness—though when used to promote some brand of meditation, app, or other contemplative "product," this inconvenient truth goes missing.

In the chapters that follow, we've used rigorous standards to sort out fluff from fact. What does science actually tell us about the impacts of meditation?

The Best We Had

4

The scene: a woodworking shop, and two fellows—we'll call them Al and Frank—are happily chatting away while Al feeds a huge sheet of plywood into the jagged blades of a giant circular saw. Suddenly you notice that Al has not used the safety guard for that saw blade—and your heartbeat speeds up as you see his thumb is headed toward that nasty sharp-toothed circle of steel.

Al and Frank are lost in their chatting, both oblivious to the danger at hand, even as that thumb heads closer to the whirring blade. Your heart races and beads of sweat form on your brow. You have the urgent wish to warn Al—but he's an actor in the film you're watching.

It Didn't Have to Happen, made by the Canadian Film Board to scare woodworkers into using their machine's safety devices, depicts three shop accidents in its twelve short minutes. Like that thumb heading inexorably into the blade, each of them builds in suspense until the moment of impact: Al loses his thumb to the circular saw;

another worker has his fingers lacerated, and a wooden plank flies into the midsection of a bystander.

The film had a life quite apart from its intended warning to wood-workers. Richard Lazarus, a psychologist at the University of California at Berkeley, deployed those depictions of gruesome accidents as a reliable emotional stressor in more than a decade of his landmark research.[1] He generously gave Dan a copy of the film to use in the research at Harvard.

Dan showed the film to some sixty people, half of them volunteers (Harvard students taking psychology courses) who had no meditation experience, the other half meditation teachers with at least two years of practice. Half the people in each group meditated just before watching the film; he taught the Harvard novices to meditate there in the lab. Dan told those assigned to a control group picked at random to simply sit and relax.

As their heart rate and sweat response jumped and subsided with the shop accidents, Dan sat in the control room next door. Experienced meditators tended to recover from the stress of seeing those upsetting events more quickly than people who were new to the practice.[2] Or so it seemed.

This research was sound enough to earn Dan a Harvard PhD and to be published in one of the top journals in his field. Even so, looking back with closer scrutiny, we see a plethora of issues and problems. Those who review grants and journal articles have strict standards for what research designs are best—that is, have the most trustworthy results. From that viewpoint, Dan's research—and the majority of studies of meditation even today—has flaws.

For instance, Dan was the person who taught the volunteers to meditate or told them to just relax. But Dan knew the desired outcome, that meditation should help more—and that could well have influenced how he spoke to the two groups, perhaps in a way that encouraged good results from meditation and poor ones from the control condition who just relaxed.

Another point: of the 313 journal articles that cited Dan's findings, not one attempted to redo the study to see if they would get similar outcomes. These authors just assumed that the results were sturdy enough to use as grounds for their own conclusions.

Dan's study is not alone; that attitude prevails still today. Replicability, as it's known in the trade, stands as a strength of the scientific method; any other scientist should be able to reproduce a given experiment and yield the same findings—or reveal the failure to reproduce them. But very, very few ever even try.

This lack of replication looms as a pervasive problem in science, particularly when it comes to studies of human behavior. While psychologists have made proposals for making psychological studies more replicable, at present little is known about how many of even the most commonly cited studies would hold up, though possibly most would.[3] And only a tiny fraction of studies in psychology are ever targets of replication; the field's incentives favor original work, not duplication. Plus, psychology, like all sciences, has a strong inbuilt publication bias: scientists rarely try to publish studies when they get no significant results. And yet that null finding itself has significance.

Then there's the crucial difference between "soft" and "hard" measures. If you ask people to report on their own behaviors, feelings, and the like—soft measures—psychological factors like a person's mood of the moment and wanting to look good or please the investigator can influence enormously how they respond. On the other hand, such biases are less (or not at all) likely to influence physiological processes like heart rate or brain activity, which makes them hard metrics.

Take Dan's research: he relied to some extent on soft measures where people evaluated their own reactions. He used a popular (among psychologists) anxiety assessment that had people rate themselves on items like "I feel worried," from "not at all" to "very much so," and from "almost never" to "almost always."[4] This method by and large showed them feeling less stressed after their first taste of meditation—a fairly common finding over the years since in meditation studies. But such self-reports are notoriously susceptible to "expectation demand," the implicit signals to report a positive outcome.

Even beginners in meditation report they feel more relaxed and less stressed once they start. Such self-reports of better stress management show up much earlier in meditators' data than do hard measures like brain activity. This could mean that the sense of lessened anxiety that meditators experience occurs before discernible shifts in the hard measures—or that the expectation of such effects biases what meditators report.

But the heart doesn't lie. Dan's study deployed physiological measures like heart rate and sweat response, which typically can't be intentionally controlled, and so yield a more accurate portrait of a person's true reactions—especially compared to those highly subjective, more easily biased self-report measures.

For his dissertation Dan's main physiological measure was the galvanic skin response, or GSR, bursts of electrical activity that signify a dollop of sweat. The GSR signals the body's stress arousal. As some speculation has it, in early evolution sweat release might have made the skin less brittle, protecting humans during hand-to-hand combat.[5]

Brain measures are even more trustworthy than "peripheral" physiological ones like heart rate. But we were too early for such methods, the least biased and most convincing of all. In the 1970s, brain imaging systems like the fMRI, SPECT, and fine-grained computerized analysis of EEG had not yet been invented.[6] Measures of responses distant in the body from the brain—heart and breath rates, sweat—were the best Dan had.[7] Because those physiological responses reflect a complex mix of forces, they are a bit messy to interpret.[8]

Another weakness of the study stems from the recording technology of the day, long before such data were digitized. Sweat rates were tracked by the sweep of a needle on a continuous spool of paper. The resulting scrawl was what Dan pored over for hours, converting ink blips into numbers for data analysis. This meant counting the smirches that signified a spurt of sweat before and after each shop accident.

The key question: Was there a meaningful difference between the four conditions—expert versus novice, told to meditate or just sit quietly—in their speed of recovery from the heights of arousal during the accidents? The results, as recorded by Dan, suggested that meditating sped up the recovery rate, and that seasoned meditators recovered quickest.[9]

That phrase *as recorded by Dan* speaks to another potential problem: it was Dan who did the scoring, and the whole endeavor was meant to support a hypothesis he endorsed. This situation favors experimenter bias, where the person designing a study and analyzing its data might skew the results toward a desired outcome.

Dan's dim (okay, very dim) recollection after nearly fifty years is that among the meditators, when there was an ambiguous GSR—one that might have been at the peak of reaction to the accident, or just afterward—he scored it as at the peak rather than at the beginning of the recovery slope. The net effect of such a bias would be to make meditators' sweat response seem to react more to the accident, while recovering more quickly (however, as we shall see, this is precisely the pattern found in the most advanced meditators studied so far).

Research on bias has found two levels: our conscious predilections and, harder to counter, our unconscious ones. To this day Dan cannot swear that his scoring of those inkspots was unbiased. Along those lines, Dan shared the dilemma of most scientists who do research on meditation: they are themselves meditators, which can encourage such bias, even if unconscious.

UNBIASING SCIENCE

It could have been a scene straight out of a Bollywood version of the *Godfather* movies: a black Cadillac limo pulled up at an assigned time and place, the back door opened, and Dan got in. Seated next to him was the big boss—not Marlon Brando/Don Corleone, but rather a smallish, bearded yogi clad in a white dhoti.

Yogi Z had come from the East to America in the 1960s and quickly captured headlines by mingling with celebrities. He attracted a huge following, and recruited hundreds of young Americans to become teachers of his method. In 1971, just before his first trip to India, Dan attended a teacher training summer camp the yogi ran.

Yogi Z somehow heard that Dan was a Harvard grad student about to travel to India on a predoctoral fellowship. The yogi had a plan for this predoc. Handing Dan a list of names and addresses of his own followers in India, Yogi Z instructed him to look each one up, interview them, and then write a doctoral dissertation with the thesis and conclusion that this particular yogi's method was the only way to become "enlightened" in this day and age.

For Dan the idea was abhorrent. Such outright hijacking of research to promote a particular brand of meditation typifies the hustle that, regrettably, has characterized a certain kind of "spiritual teacher" (remember Swami X). When such a teacher engages in the self-promotion typical of some commercial brand, it signals that someone hopes to use the appearance of inner progress in the service of marketing. And when researchers wed to a particular brand of meditation report positive findings, the same questionable bias arises, as well as another question: Were there negative results that went unreported?

For instance, the meditation teachers in Dan's study taught Transcendental Meditation (TM). TM research has had a somewhat checkered history in part because most of it has been done by staff at Maharishi University of Management (formerly Maharishi International University), which is a part of the organization that promotes TM. This raises the concern of a conflict of interest, even when the research has been well done.

For this reason, Richie's lab intentionally employs several scientists who are skeptical of meditation's effects, and who raise a healthy number of issues and questions that "true believers" in the practice might overlook or sweep under the rug. One result: Richie's lab has published several nonfindings, studies that test a specific hypothesis about the effect of meditation and fail to observe the expected effect. The lab also publishes failures to replicate—studies that do not get the same results when duplicating the method of previously published papers that found meditation has some beneficial effect. Such failures to replicate earlier findings call them into question.

Bringing in skeptics is but one of many ways to minimize

experimenter bias. Another would be to study a group that is told about meditation practices and their benefits but gets no instruction. Better: an "active control," where one group engages in an activity unlike meditation, one that they believe will benefit them, such as exercise.

A further dilemma in our Harvard research, also still pervasive in psychology, was that the undergrads available for study in our lab were not typical of humanity as a whole. Our experiments were done with subjects known in the field as WEIRD: Western, educated, industrialized, rich, and from democratic cultures.[10] And using Harvard students, an outlier group even among the WEIRD, makes the data less valuable in searching for universals in human nature.

THE VARIETIES OF THE MEDITATIVE EXPERIENCE

Richie in his dissertation research was among the first neuroscientists to ask if we can identify a neural signature of attention skill. That basic question was, in those days, quite respectable.

But Richie's PhD research was in the spirit of that concealed excursion into the mind in his undergraduate work. The agenda embedded, *sub rosa,* in the study: exploring if signs of skill in attention differed in meditators and nonmeditators. Did meditators get better at focusing? In those days, that was *not* a respectable question.

Richie measured the brain electrical signals from the scalp of meditators as they heard tones or saw flashing LED lights, while he instructed them to focus on the sounds and ignore the lights, or vice versa. Richie analyzed the electrical signals for "event-related potential"

(ERP), indicated by specific blips in response to a light and/or tone. The ERP, embedded in a chorus of noise, is a signal so minuscule it is measured in microvolts—millionths of a volt. These tiny signals offer a window on how we allocate our attention.

Richie found that the size of these tiny signals was diminished in response to the tone when meditators focused on the light, while the signals triggered by the light were reduced in size when the meditators focused their attention on the tone. That finding alone would be ho-hum; we would expect that. But this pattern of blocking out the unwanted modality was much stronger in the meditators than in the controls—some of the first evidence that meditators were better at focusing their attention than nonmeditators.

Since selecting a target for focus and ignoring distractions marks a key attention skill, Richie concluded that brain electrical recordings—the EEG—could be used for this assessment (routine today, but a step in scientific progress back then). Still, the evidence that meditators were any better at this than the control group, who had never meditated, was rather weak.

In retrospect, we can see one reason why this evidence was in itself questionable: Richie had recruited a mix of meditators, who deployed various methods. Back in 1975 we were quite naive about how important these variations in technique were. Today we know there are many aspects of attention, and that different kinds of meditation train a variety of mental habits, and so, impact mental skills in varying ways.

For example, researchers at the Max Planck Institute for Human Cognitive and Brain Sciences in Leipzig, Germany, had novices practice daily for a few months three different types of meditation: focusing on breathing; generating loving-kindness; and monitoring thoughts without getting swept away by them.[11] Breath focus, they found, was calming—seeming to confirm a widespread assumption about meditation's usefulness as a means to relax. But in contradiction to that stereotype, neither the loving-kindness practice nor monitoring thoughts made the body more relaxed, apparently because each demands mental effort: for example, while watching thoughts you continually get swept up in them—and then, when you notice this has happened, need to make a conscious effort to simply watch again. In addition, the loving-kindness practice, where you wish yourself and others well, understandably created a positive mood, while the other two methods did not.

So, differing types of meditation produce unique results—a fact that should make it a routine move to identify the specific type being studied. Yet confusion about the specifics remains all too common. One research group, for instance, has collected state-of-the-art data on brain anatomy in fifty meditators, an invaluable data set.[12] Except that the names of the meditation practices being studied reveals a mixture of types—a hodgepodge. Had the specific mental training entailed by each meditation type been methodically recorded, that data set might well yield even more valuable findings. (Even so, kudos for disclosing this information, which too often goes unnoted.)

As we read through the now vast trove of research on meditation, we sometimes wince when we come across the confusion and naiveté of some scientists about the specifics. Too often they are simply mistaken, like the scientific article that said that in both Zen and Goenka-style vipassana, meditators have their eyes open (what's wrong here: Goenka has people close their eyes).

A handful of studies have used an "antimeditation" method as an active control. In one version of this so-called antimeditation, volunteers were told to concentrate on as many positive thoughts as possible—which actually resembles some contemplative methods, such as the loving-kindness meditation we will review in chapter six. The fact that those experimenters thought this was unlike meditation speaks to their confusion about what exactly they were researching.

The rule of thumb—that what gets practiced gets improved—underscores the importance of matching a given mental strategy in meditation to its result. This is true equally for those who study meditation and those who meditate: one must be aware of the likely outcomes from a given meditation approach. They are not all the same, contrary to the misunderstanding among some researchers, and even practitioners.

In the realm of mind (as everywhere else), what you do determines what you get. In sum, "meditation" is not a single activity but a wide range of practices, all acting in their own particular ways in the mind and brain.

Lost in Wonderland, Alice asked the Cheshire Cat, "Which way should I go?"

He replied, "That depends on where you want to get to."

The Cheshire Cat's advice to Alice holds, too, for meditation.

COUNTING THE HOURS

Each of Dan's "expert" meditators, all Transcendental Meditation teachers, had practiced TM for at least two years. But Dan had no way of knowing how many total hours they had put in over those years. Nor did he know what the actual quality of those hours might have been.

Few researchers, even today, have this crucial piece of data. But, as we will see in more detail in chapter thirteen, "Altering Traits," our model of change tracks how many lifetime hours of practice a meditator has done and whether it was daily or on retreat. These total hours are then connected with shifts in qualities of being and the underlying differences in the brain that give rise to them.

Very often meditators are lumped into gross categories of experience, like "beginner" and "expert," without any further specifics. One research group reported the daily time the people they studied put into meditation—ranging from ten minutes a few times a week to 240 minutes daily—but not how many months or years they had done so, which is essential in calculating lifetime hours of practice.

Yet this calculation goes missing in the vast majority of meditation studies. So that classic Zen study from the 1960s showing a failure to habituate to repeated sounds—one of the few existing then and one that had gotten us interested in the first place—actually gave sparse data on the Zen monks' meditation experience. Was it an hour a day, ten minutes, zero on some days, or six hours every day? How many retreats (*sesshins*) of more intensive practice did they do, and how many hours of meditation did each involve? We have no idea.

To this day the list of studies that suffer from this uncertainty could go on and on. But getting detailed information about the total lifetime hours of a meditator's practice has become standard operating procedure in Richie's lab. Each of the meditators they study report on what kind of meditation practice they do, how often and for how long they do it in a given week, and whether they go on retreats.

If so, they note how many hours a day they practice on retreat, how long the retreat is, and how many such retreats they have done. Even further, the meditators carefully review each retreat and estimate the time spent doing different styles of meditation practice. This math allows the Davidson group to analyze their data in terms of total hours of practice and separate the time for different styles and for retreat versus home hours.

As we will see, there sometimes is a dose-response relationship when it comes to the brain and behavioral benefits from meditation: the more you do it, the better the payoff. That means that when researchers fail to report the lifetime hours of the meditators they are studying, something important has gone missing. By the same token, too many meditation studies that include an "expert" group show wild variation in what that term means—and don't use a precise metric for how many hours those "experts" have practiced.

If the people being studied are meditating for the first time—say, being trained in mindfulness—their number of practice hours is straightforward (the instruction hours plus however many they do at home on their own). Yet many of the more interesting studies look at seasoned meditators without calculating each person's lifetime hours, which can vary greatly. One, for example, lumped together meditators who had from one year of experience to twenty-nine years!

Then there's the matter of expertise among those giving meditation instruction. A handful of studies among the many we looked at thought to mention how many years of experience in meditation the teachers had, though none calculated their lifetime hours. In one study the upper number was about fifteen years; the lowest was zero.

BEYOND THE HAWTHORNE EFFECT

Way back in the 1920s, at the Hawthorne Works, a factory for electrical equipment near Chicago, experimenters simply improved lighting in that factory and slightly adjusted work schedules. But, with even those small changes for the better, people worked harder—at least for a while.

The take-home: any positive intervention (and, perhaps, simply having someone observe your behavior) will move people to say they feel better or improve in some other way. Such "Hawthorne effects," though, do not mean there was any unique value-added factor from a given intervention; the same upward bump would occur from any change people regarded as positive.

Richie's lab, sensitized to issues like the Hawthorne effect, has devoted considerable thought and effort to using proper comparison conditions in their studies of meditation. The instructor's enthusiasm for a given method can infect those who learn it—and so the "control" method should be taught with the same level of positivity as is true for the meditation.

To tease out extraneous effects like these from the actual impacts of meditation, Richie and his colleagues developed a Health Enhancement Program (HEP) as a comparison condition for studies of mindfulness-based stress reduction. HEP consists of music therapy with relaxation; nutritional education; and movement exercises like posture improvement, balance, core strengthening, stretching, and walking or jogging.

In the labs' studies, the instructors who taught HEP believed it would help, just as much as did those who taught meditation. Such an "active control" can neutralize factors like enthusiasm, and so better identify the unique benefits of any intervention—in this case, meditation—to see what it adds over and above the Hawthorne edge.

Richie's group randomly assigned volunteers to either HEP or mindfulness-based stress reduction (MBSR) and then before and after the training had them fill out questionnaires that in earlier research had reflected improvements from meditation. But in this study, both groups reported comparable improvement on these subjective measures of general distress, anxiety, and medical symptoms. This led Richie's group to conclude that much of the stress relief improvements beginners credit to meditation do not seem to be that unique.[13]

Moreover, on a questionnaire that was specifically developed to measure mindfulness, absolutely no difference was found in the level of improvement from MBSR or HEP.[14]

This led Richie's lab to conclude that for this variety of mindfulness, and likely for any other meditation, many of the reported benefits in the early stages of practice can be chalked up to expectation, social bonding in the group, instructor enthusiasm, or other "demand characteristics." Rather than being from meditation per se, any reported benefits may simply be signs that people have positive hopes and expectations.

Such data are a warning to anyone looking for a meditation practice to be wary of exaggerated claims about its benefits. And also a wake-up call to the scientific community to be more rigorous in designing meditation studies. Just finding that people practicing one or another kind of meditation report improvements compared to those in a control group who do nothing does not mean such benefits are due to the meditation itself. Yet this is perhaps the most common paradigm still used in research on the benefits of meditation—and it clouds the picture of what the true advantages of the practice might be.

We might expect similar enthusiastic reports from someone who expects a boost in well-being by taking up Pilates, bowling, or the Paleo Diet.

WHAT EXACTLY *IS* "MINDFULNESS"?

Then there is the confusion about what we mean by *mindfulness*, perhaps the most popular method du jour among researchers. Some scientists use the term as a stand-in for any and all kinds of meditation. And in popular usage, mindfulness can refer to meditation in general, despite the fact that mindfulness is but one of a wide variety of methods.

To dig down a bit, *mindfulness* has become the most common English translation of the Pali language's word *sati*. Scholars, however, translate *sati* in many other ways—"awareness," "attention," "retention," even "discernment."[15] In short, there is not a single English equivalent for *sati* on which all experts agree.[16]

Some meditation traditions reserve "mindfulness" for noticing when the mind wanders. In this sense, mindfulness becomes part of a larger sequence which starts with a focus on one thing, then the mind wandering off to something else, and then the mindful moment: noticing the mind has wandered. The sequence ends with returning attention to the point of focus.

That sequence—familiar to any meditator—could also be called "concentration," where mindfulness plays a supporting role in the effort to focus on one thing. In one-pointed focus on a mantra, for example, sometimes the instruction is, "Whenever you notice your mind wandering, gently start the mantra again." In the mechanics of meditation, focusing on one thing only means also noticing when your mind wanders off so you can bring it back—and so concentration and mindfulness go hand in hand.

Another common meaning of *mindfulness* refers to a floating awareness that witnesses whatever happens in our experience without judging or otherwise reacting. Perhaps the most widely quoted definition comes from Jon Kabat-Zinn: "The awareness that emerges through paying attention on purpose, in the present moment, and nonjudgmentally to the unfolding of experience."[17]

From the viewpoint of cognitive science, there's another twist when it comes to the precise methods used: what's called "mindfulness," by scientists and practitioners alike, can refer to very different ways to deploy attention. For example, the way mindfulness gets defined in a Zen or Theravadan context looks little like the understanding of the term in some Tibetan traditions.

Each refers to differing (sometimes subtly so) attentional stances—and quite possibly to disparate brain correlates. So it becomes essential that researchers understand what kind of mindfulness they are actually studying—or if, indeed, a particular variety of meditation actually *is* mindfulness.

The meaning of the term *mindfulness* in scientific research has taken a strange turn. One of the most commonly used measures of mindfulness was not developed on the basis of what happens during actual mindfulness meditation but rather by testing hundreds of college undergraduates on a questionnaire that the researchers thought would capture different facets of mindfulness.[18] For example, you are asked whether statements like these are true for you: "I watch my feelings without getting carried away by them" or "I find it difficult to stay focused on what's happening in the present moment."

The test includes qualities like not judging yourself—for example, when you have an inappropriate feeling. This all seems fine at first glance. Such a measure of mindfulness should and does correlate with people's progress in training programs like MBSR, and the test scores correlated with the amount and quality of mindfulness practice itself.[19] From a technical viewpoint that's very good—it's called "construct validity" in the testing trade.

But when Richie's group put that measure to another technical test, they found problems in "discriminant validity," the ability of a measure not just to correlate with what it should—like MBSR—but also *not* to correlate when it should not. In this case, that test should not reflect the changes among those in the HEP active control group, which was intentionally designed *not* to enhance mindfulness in any way.

But the results from the HEP folks were pretty much like those from MBSR—an uptick in mindfulness as assessed on the self-report test. More formally, there was zero evidence that this measure had discriminant validity. *Oops.*

Another widely used self-report measure of mindfulness, in one study, showed a positive correlation between binge drinking and mindfulness—the more drinking, the greater the mindfulness. Seems like something is off-base here![20] And a small study with twelve seasoned (average of 5,800 hours of practice) and twelve more expert meditators (average of 11,000 hours of practice) found they did not differ from a nonmeditating group on two very commonly used questionnaire measures of mindfulness, perhaps because they are *more* aware of the wanderings of their mind than most people.[21]

Any questionnaire that asks people to report on themselves can be susceptible to skews. One researcher put it more bluntly: "These can be gamed." For that reason the Davidson group has come up with what they consider a more robust behavioral measure: your ability to maintain focus as you count your breaths, one by one.

This is not as simple as it may sound. In the test you press the down arrow on a keyboard on each outbreath. And to up the odds, on every ninth exhale you tap a different key, the right arrow. Then you start counting your breaths from one to nine again.[22] The strength of this test: the difference between your count and the actual number of breaths you took renders an objective measure far less prone to psychological bias. When your mind wanders, your counting accuracy suffers. As expected, expert meditators perform significantly better than nonmeditators, and scores on this test improve with training in mindfulness.[23]

All of this cautionary review—the troubles with our first attempts at meditation research, the advantages of an active control group, the need for more rigor and precision in measuring meditation impacts—seems a fitting prelude to our wading into the rising sea of research on meditation.

In summarizing these results we've tried to apply the strictest experimental standards, which lets us focus on the very strongest findings. This means setting aside the vast majority of research in meditation—including results scientists view as questionable, inconclusive, or otherwise marred.

As we've seen, our somewhat flawed research methods during our Harvard graduate school days reflected the general quality—or lack of it—during the first decades of meditation studies, the 1970s and 1980s. Today our initial research attempts would not meet our own standards to be included here. Indeed, a large proportion of meditation studies in one way or another fail to meet the gold standards for research methods that are essential for publication in the top, "A-level" scientific journals.

To be sure, over the years there has been a ratcheting upward of sophistication as the number of studies of meditation has exploded to more than one thousand per year. This tsunami of meditation research creates a foggy picture, with a confusing welter of results. Beyond our focus on the strongest findings, we try to highlight the meaningful patterns within that chaos.

We've broken down this mass of findings along the lines of trait changes described in the classic literature of many great spiritual traditions. We see such texts as offering working hypotheses from ancient times for today's research.

We've also related these trait changes to the brain systems involved, wherever the data allow. The four main neural pathways meditation transforms are, first, those for reacting to disturbing events—stress and our recovery from it (which Dan tried not so successfully to document). As we will see, the second brain system, for compassion and empathy, turns out to be remarkably ready for an upgrade. The third, circuitry for attention, Richie's early interest, also improves in several ways—no surprise, given that meditation at its core retrains our habits of focus. The fourth neural system, for our very sense of self, gets little press in modern talk about meditation, though it has traditionally been a major target for alteration.

When these strands of change are twined together, there are two major ways anyone benefits from contemplative effort: having a healthy body and a healthy mind. We devote chapters to the research on each of these.

In teasing out the main trait effects of meditation, we faced a gargantuan task—one that we've simplified by limiting our conclusions to the very best studies. This more rigorous look contrasts with the too-common practice of accepting findings—and touting them—simply because they are published in a "peer-reviewed" journal. For one, academic journals themselves vary in the standards by which peers review articles; we've favored A-level journals, those with the highest standards. For another, we've looked carefully at the methods used, rather than ignoring the many drawbacks and limitations to these published studies that are dutifully listed at the ends of such articles.

To start, Richie's research group gathered an exhaustive collection for a given topic like compassion from all journal articles published on the effects of meditation. They then winnowed them to select those that met the highest standards of experimental design. So, for example, of the original 231 reports on cultivating loving-kindness or compassion, only 37 met top design standards. And when Richie looked through the lenses of design strength and of importance, eliminated overlap, and otherwise distilled them, this closer scrutiny shrank that number to 8 or so studies, whose findings we talk about in chapter six, "Primed for Love," along with a few others that raise compelling issues.

Our scientific colleagues might expect a far more detailed—okay, obsessive—accounting of all the relevant studies, but that's not our agenda here. That said, we should nod with great appreciation to the many research efforts we did not include whose findings agree with our account (or disagree, or add a twist), some excellent and some not so.

But let's keep this simple.

A Mind Undisturbed

5

E verything you do, be it great or small, is but one-eighth of the
problem," a sixth-century Christian monk admonished his fellow
renunciates, "whereas to keep one's state undisturbed even if thereby
one should fail to accomplish the task, is the other seven-eighths."[1]

A mind undisturbed marks a prominent goal of meditation paths
in all the great spiritual traditions. Thomas Merton, a Trappist contemplative, wrote his own version of a poem lauding this very quality,
taken from the ancient annals of Taoism. He tells of a craftsman who
could draw perfect circles without using a compass, and whose mind
was "free and without concern."[2]

A mind unworried has as its opposite the angst life brings us:
money worries, working too hard, family problems, health troubles. In
nature, stress episodes like encountering a predator are temporary,
giving the body time to recover. In modern life stressors are mostly
psychological, not biological, and can be ongoing (if only in our

thoughts), like a horrific boss or trouble with family. Such stressors trigger those same ancient biological reactions. If these stress reactions last for a long time, they can make us sick.

Our vulnerability to stress-worsened diseases like diabetes or hypertension reflects the downside in our brain's design. The upside reflects the glories of the human cortex, which has built civilizations (and the computer this is being written on). But the brain's executive center, located behind the forehead in our prefrontal cortex, gives us both a unique advantage among all animals and a paradoxical disadvantage: the ability to anticipate the future—and worry about it—as well as to think about the past—and regret.

As Epictetus, a Greek philosopher, put it centuries ago, it's not the things that happen to us that are upsetting but the view we take of those doings. A more modern sentiment comes from poet Charles Bukowski: it's not the big things that drive us mad, but "the shoelace that snaps with no time left."

The science here shows that the more we perceive such hassles in our lives, the higher our levels of stress hormones like cortisol. That's a bit ominous: cortisol, if raised chronically, has deleterious impacts like an increased risk of dying from heart disease.[3] Can meditation help?

FROM THE BACK OF AN ENVELOPE

We first got to know Jon Kabat-Zinn during our Harvard days, when he had just finished his PhD in molecular biology at MIT and was exploring meditation and yoga. Jon was a student of Korean Zen master Seung Sahn, who had a meditation center in the same Cambridge neighborhood where Dan was living. And not far away, in Richie's second-floor apartment off Harvard Square, Jon gave Richie his first instruction in meditation and yoga, shortly before Richie's trip to India.

A like-minded meditating scientist, Jon had joined our team when we studied Swami X at Harvard Medical School. Jon had just gotten a fellowship in anatomy and cell biology at the newly opened University of Massachusetts Medical School in Worcester, an hour's drive from Cambridge. The anatomy was what interested him most—Jon had already begun teaching yoga classes in Cambridge.

In those days Jon often went on retreats at the Insight Meditation Society (IMS), then recently founded, in Barre, also about an hour away from Boston and not far from Worcester. In 1974, several years before IMS was founded, Jon had spent two weeks one freezing early April in an unheated Girl Scout camp in the Berkshires, rented for a vipassana course. The teacher, Robert Hover, had been commissioned to teach by the Burmese master U Ba Khin, who, you might remember, was also the teacher of S. N. Goenka, whose retreats Dan and Richie attended in India.

Like Goenka, the main methods Hover taught were, initially, to focus on your breath in order to build concentration for the first three days of the retreat, and then to systematically scan the body's sensations very slowly, from head to toe, over and over again for the next seven days. During the scan you focused only on the bare bodily sensations—the norm in that meditation lineage.

Hover's instructions included several two-hour meditation sittings during which students vowed not to make a single voluntary movement—twice as long as those at Goenka's courses. These immobile sessions produced a level of pain, Jon said, he had never experienced in his life. But as he sat through that unbearable pain and scanned his body to focus on his experience, the pain dissolved into pure sensations.

On this retreat Jon had an insight, which he quickly wrote down on the back of an envelope, that there might be a way to share the benefits of meditation practices with medical patients, especially those experiencing chronic pain that wouldn't go away just by changing their posture or stopping the meditation practice. Coupled with a sudden vision that came to him a few years later on a retreat at IMS and that drew together disparate parts of his own practice history into a form that would be accessible to anyone, the program now known around the world as mindfulness-based stress reduction, or MBSR, came into being in September of 1979 at the University of Massachusetts Medical Center.[4]

In his vision he realized that pain clinics are filled with people whose symptoms are excruciating and who can't escape the pain except through debilitating narcotics. He saw that the body scan and other mindfulness practices could help these patients uncouple the cognitive and emotional parts of their experience of pain from the pure sensation, a perceptual shift that can itself be a significant relief.

But most of these patients—a random slice of folks from the working-class environs of Worcester—could not sit still for long periods of time like the dedicated meditators Hover taught. So Jon adapted a method from his yoga training, a lying-down body scan meditation which, similar to the Hover approach, has you connect with and then move through key regions of the body in a systematic sequence, starting with the toes of the left foot, and winding up at the top of the head. The key point: it is possible to register and then investigate and transform your relationship to whatever you are sensing at a given place in the body, even if it is highly unpleasant.

Borrowing from both his Zen background and vipassana, Jon added a sitting meditation where people pay careful attention to their breath, letting go of thoughts or sensations that arise—just being aware of attending itself, not of the object of attention, the breath at the beginning, and then other objects such as sounds, thoughts, emotions, and of course, bodily sensations of all kinds. And, taking another cue from Zen and vipassana, he added mindful walking, mindful eating, and a general awareness of life's activities, including one's relationships.

We were pleased that Jon pointed to our Harvard research as evidence (otherwise pretty scant in those days) that methods taken from contemplative paths and put in new forms without their spiritual context could have benefits in the modern world.[5] These days that evidence has grown more than ample; MBSR has risen to the top of meditation practices undergoing scientific scrutiny. MBSR may be the most widely practiced form of mindfulness anywhere, taught around the world in hospitals and clinics, schools, even businesses. One of the many benefits claimed for MBSR: boosting how well people handle stress.

In an early study of the impact of MBSR on stress reactivity, Philippe Goldin (an SRI attendee) and his mentor at Stanford University, James Gross, studied a small group of patients with social anxiety disorder who underwent the standard eight-week MBSR program.[6] Before and after the training, they went into the fMRI scanner, while being presented with stressors—statements taken from their own tales of social "meltdowns" and their thoughts during them—for example, "I am incompetent," or "I am ashamed of my shyness."

As these stressful thoughts were presented, the patients used either of two different attentional stances: mindful awareness of their breath or distraction by doing mental arithmetic. Only mindfulness of their breath both lowered activity in the amygdala—mainly via a faster recovery—and strengthened it in the brain's attentional networks, while the patients reported less stress reactivity. The same beneficial pattern emerged when the patients who had done MBSR were compared with some who had trained in aerobics.[7]

That is but one of many hundreds of studies that have been done on MBSR, revealing a multitude of payoffs, as we'll see throughout this book. But the same can be said for MBSR's close cousin, mindfulness itself.

MINDFUL ATTENTION

When we started to participate in dialogues between the Dalai Lama and scientists at the Mind and Life Institute, we were impressed by the precision with which one of his interpreters, Alan Wallace, was able to equate scientific terms with their equivalent meanings in Tibetan, a language lacking any such technical terminology. Alan, it turned out, had a PhD in religious studies from Stanford University, extensive familiarity with quantum physics, and rigorous philosophical training, in part as a Tibetan Buddhist monk for several years.

Drawing on his contemplative expertise, Alan developed a unique program that extracts from the Tibetan context a meditation practice accessible to anyone, what he calls Mindful Attention Training. This program starts with full focus on the breath, then progressively refines attention to observe the natural flow of the mind stream and finally rest in the subtle awareness of awareness itself.[8]

In a study at Emory, people who had never meditated previously were randomly assigned to practice Mindful Attention Training or a compassion meditation. A third group, an active control, went through a series of discussions on health.[9]

The participants were scanned before and after they underwent eight weeks of training. While in the scanner they viewed a set of images—standard in emotion research—which includes a few upsetting ones, such as a burn victim. The Mindful Attention group showed reduced amygdala activity in response to the disturbing pictures. The changes in amygdala function occurred in the ordinary baseline state in this study, suggesting the seeds of a trait effect.

A word about the amygdala, which has a privileged role as the brain's radar for threat: it receives immediate input from our senses, which it scans for safety or danger. If it perceives a threat, the amygdala circuitry triggers the brain's freeze-fight-or-flight response, a stream of hormones like cortisol and adrenaline that mobilize us for action. The amygdala also responds to anything important to pay attention to, whether we like or dislike it.

The sweat dollops Dan measured in his study were distant indicators of this amygdala-driven reaction. In effect, Dan was trying to tease out a change in amygdala function—a quicker recovery from arousal—but was using a hopelessly indirect metric with the sweat response. That was in a day long before the invention of scanners that directly track activity in brain regions.

The amygdala connects strongly to brain circuitry for both focusing our attention and for intense emotional reactions. This dual role explains why, when we are in the grip of anxiety, we are also very distracted, especially by whatever is making us anxious. As the brain's radar for threat, the amygdala rivets our attention on what it finds troubling. So when something worries or upsets us, our mind wanders over and over to that thing, even to the point of fixation—like the viewers of the shop accident film when they saw Al's thumb approach that wicked saw blade.

About the same time as Alan's findings that mindfulness calms the amygdala, other researchers had volunteers who had never meditated before practice mindfulness for just twenty minutes a day over one week, and then have an fMRI scan.[10] During the scan they saw images ranging from gruesome burn victims to cute bunnies. They watched these images in their everyday state of mind, and then while practicing mindfulness.

During mindful attention their amygdala response was significantly lower (compared to nonmeditators) to all the images. This sign of being less disturbed, tellingly, was greatest in the amygdala on the brain's right side (there are amygdalae in both right and left hemispheres), which often has a stronger response to whatever upsets us than the one on the left.

In this second study, lessened amygdala reactivity was found only during mindful attention and not during ordinary awareness, indicating a state effect, not an altered trait. A trait change, remember, is the "before," not the "after."

PAIN IS IN THE BRAIN

If you give the back of your hand a hard pinch, different brain systems mobilize, some for the pure sensation of pain and others for our dislike of that pain. The brain unifies them into a visceral, instant *Ouch!*

But that unity falls apart when we practice mindfulness of the body, spending hours noticing our bodily sensations in great detail. As we sustain this focus, our awareness morphs.

What had been a painful pinch transforms, breaking down into its constituents: the intensity of the pinch and the painful sensation, and the emotional feeling tone—we don't want the pain; we urgently want the pain to stop.

But if we persevere with mindful investigation, that pinch becomes an experience to unpack with interest, even equanimity. We can see our aversion fall away, and the "pain" break down into subtler flavors: throbbing, heat, intensity.

Imagine now you hear a soft rumble as a five-gallon tank of water starts boiling and sends a stream of fluid through the thin rubber hose that runs through the two-inch square metal plate strapped tight on your wrist. The plate heats up, pleasantly at first. But that pleasantness quickly heads toward pain, as the water temperature jumps several degrees within a couple of seconds. Finally, you can't take it anymore—if this were a hot stove you had touched, you would instantly pull away. But you can't remove that metal plate. You feel the almost excruciating heat for a full ten seconds, sure you are getting burned.

But you get no burn; your skin is fine. You've just reached your highest pain threshold, exactly what this device, the Medoc thermal stimulator, was designed to detect. Used by neurologists to assess conditions like neuropathy that reveal deterioration of the central nervous system, the thermal stimulator has built-in safety devices so people's skin won't be burned, even as it calibrates precisely their maximum pain threshold. And people's pain thresholds are nowhere near the higher range at which burns occur. That's why the Medoc has been used with experimental volunteers to establish how meditation alters our perceptions of pain.

Among pain's main components are our purely physiological sensations, like burning, and our psychological reactions to those sensations.[11] Meditation, the theory goes, might mute our emotional response to pain and so make the heat sensations more bearable.

In Zen, for example, practitioners learn to suspend their mental reactions and categorization of whatever arises in their minds or around them, and this mental stance gradually spills over into everyday life.[12] "The experienced practitioner of zazen does not depend on sitting quietly," as Ruth Sasaki, a Zen teacher, put it, adding, "States of consciousness at first attained only in the meditation hall gradually become continuous in any and all activities."[13]

Seasoned Zen meditators who were having their brains scanned (and who were asked to "not meditate") endured the thermal stimulator.[14] While we've noted the reasons to have an active control group, this research had none. But that's less an issue here, because of the brain imaging. If the outcome measures are based on self-reports (the most easily swayed by expectations) or even behavior observed by someone else (somewhat less susceptible to bias) then an active control group matters greatly. But when it comes to their brain activity, people have no clue what's going on, and so an active control matters less.

The more experienced among the Zen students not only were able to bear more pain than could controls, they also displayed little activity in executive, evaluative, and emotion areas during the pain—all regions that ordinarily flare into activity when we are under such intense stress. Tellingly, their brains seemed to disconnect the usual link between executive center circuits where we evaluate (*This hurts!*) and circuitry for sensing physical pain (*This burns*).

In short, the Zen meditators seemed to respond to pain as though it was a more neutral sensation. In more technical language, their brains showed a "functional decoupling" of the higher and lower brain regions that register pain—while their sensory circuitry felt the pain, their thoughts and emotions did not react to it. This offers a new twist on a strategy sometimes used in cognitive therapy: *reappraisal* of severe stress—thinking about it in a less threatening way—which can lessen its subjective severity as well as the brain's response. Here, though, the Zen meditators seemed to apply a *no-appraisal* neural strategy—in keeping with the mind-set of zazen itself.

A close reading of this article reveals a mention only in passing of a significant trait effect, in a difference found between Zen meditators and the comparison group. During the initial baseline reading the temperature is increased in a staircase-like series of finely graduated rises to calibrate the precise maximum pain threshold for each person. The Zen practitioners' pain threshold was 2 degrees Centigrade (5.6 degrees Fahrenheit) higher than for nonmeditators.

This may not sound like much, but the way we experience pain from heat means that slight increases in temperature can have dramatic impact both subjectively and in how our brain responds. Though that difference of 2 degrees Centigrade may seem trivial, in the world of pain experience, it is huge.

Researchers are, appropriately, skeptical about such traitlike findings because self-selection in who chooses to stick with meditation and who drops out along the way might also account for such data; perhaps people who choose to meditate for years and years are already different in ways that look like trait effects. The maxim "Correlation does not mean causation" applies here.

But if a trait can be understood as a lasting effect of the practice, that poses an alternative explanation. And when different research groups come up with similar trait findings, these converging results make us take the result more seriously.

Contrast the Zen sitters' recovery from stress reactivity with burnout, the depleted, hopeless state that comes from years of constant, unremitting pressures, like from jobs that demand too much. Burnout has become rampant among health care professions such as nurses and doctors, as well as those who care at home for loved ones with problems like Alzheimer's. And, of course, anyone can feel burned-out who faces the rants of rude customers or continual implacable deadlines, as with the hectic pace of a business start-up.

Such constant stress sculpts the brain for the worse, it seems.[15] Brain scans of people who for years had faced work that demanded up to seventy hours each week revealed enlarged amygdalae and weak connections between areas in the prefrontal cortex that can quiet the amygdala in a disturbing moment. And when those stressed-out workers were asked to reduce their emotional reaction to upsetting pictures, they were unable to do so—technically, a failure in "down-regulation."

Like people who suffer from post-traumatic stress syndrome, victims of burnout are no longer able to put a halt to their brain's stress response—and so, never have the healing balm of recovery time.

There are tantalizing results that indirectly support meditation's role in resilience. A collaboration between Richie's lab and the research group directed by Carol Ryff looked at a subset of participants in a large, multisite, national study of midlife in the United States. They found that the stronger a person's sense of purpose in life, the more quickly they recovered from a lab stressor.[16]

Having a sense of purpose and meaning may let people meet life's challenges better, reframing them in ways that allow them to recover more readily. And, as we saw in chapter three, meditation seems to enhance well-being on Ryff's measure, which includes a person's sense of purpose. So what's the direct evidence that meditation can help us meet upsets and challenges with more aplomb?

BEYOND CORRELATION

When Dan taught the psychology of consciousness course in 1975 at Harvard, Richie, then in his last year of graduate school, was, as mentioned, a teaching assistant. Among the students he met with weekly was Cliff Saron, then a senior at Harvard. Cliff had a knack for the technical end of research, including the electronics (perhaps a legacy of his father, Bob Saron, who had managed the sound equipment at NBC). Cliff's adeptness soon made him a coauthor on research papers with Richie.

And when Richie got his first teaching post at the State University of New York at Purchase, he took Cliff along to manage the laboratory. After a stint there—and coauthoring a slew of scientific papers with Richie—Cliff got his own PhD in neuroscience at Albert Einstein College of Medicine. He now directs a lab at the Center for Mind and Brain at the University of California at Davis, and has often been on the faculty at the Mind and Life Summer Research Institute.

Cliff's astute sense of methodological issues no doubt helped him design and run a crucial bit of research, one of the few longitudinal studies of meditation to date.[17] With Alan Wallace as retreat leader, Cliff put together a rigorous battery of assessments for students going through a three-month training in a range of classic meditation styles, including some, like mindfulness of breathing, meant to increase focus and others to cultivate positive states like loving-kindness and equanimity. While the "yogis" pursued their demanding schedule of meditating six or more hours a day for ninety days, Cliff had them take a battery of tests at the beginning, middle, and end of the retreat, and five months after the retreat had concluded.[18]

The comparison group was people who had signed up for the three-month retreat but who did not start until the first group finished. Such a "wait-list" control eliminates worries about expectation demand and similar psychological confounds (but does not add an active control like HEP—which would be a logistic and financial burden in a study like this). A stickler for precision in research, Cliff flew people in the wait-list group to the retreat place and gave them exactly the same assessments in the identical context as those in the retreat.

One test presented lines of different lengths in rapid succession, with the instruction to press one button for a line that was shorter than the others. Only one out of ten lines was short; the challenge is to inhibit the knee-jerk tendency to press the button for a short line when a long one appears. As the retreat progressed, so did the ability of the meditators to control this impulse—a mirror on a skill critical to managing our emotion, the capacity to refrain from acting on whim or impulse.

This simple skill, statistical analyses suggested, led to a range of improvements on self-reports, from less anxiety to an overall sense of well-being, including emotion regulation as gauged by reports of recovering more quickly from upsets and more freedom from impulses. Tellingly, the wait-list controls showed no change in any of these measures—but showed the same improvements once they had gone through the retreat.

Cliff's study directly ties these benefits to meditation, lending strong support to the case for altered traits. A clincher: a follow-up five months after the retreats ended found that the improvements remained.

And the study dispels doubts that all the positive traits found in long-term meditators are simply due to self-selection, where people who already had those traits choose the practice or stay with it in the long run. From evidence like this, it seems likely that the states we practice in meditation gradually spill over into daily life to mold our traits—at least when it comes to handling stress.

A DEVILISH ORDEAL

Imagine you are describing your qualifications for a job while two interviewers glare at you, unsmiling. Their faces reveal no empathy, not even an encouraging nod. That's the situation in the Trier Social Stress Test (TSST), one of the most reliable ways known to science to trigger the brain's stress circuits and its cascade of stress hormones.

Now imagine, after that dispiriting job interview, doing some pressured mental arithmetic: you have to subtract 13s in rapid-fire succession from a number like 1,232. That's the second part of the Trier test, and those same impassive interviewers push you to do the math faster and faster—and whenever you make a mistake, they tell you to start all over at 1,232. That devilish test delivers a huge dose of social stress, the awful feelings we get when other people evaluate, reject, or exclude us.

Alan Wallace and Paul Ekman created a renewal program for schoolteachers that combined psychological training with meditation.[19] Whereas Dan had used the shop accident film to bring stress into the lab, here the stressor was the Trier test's simulated job interview followed by that formidable math challenge.

The more hours those teachers had practiced meditation, the quicker their blood pressure recovered from a high point during the TSST. This was true five months after the program ended, suggesting at least a mild trait effect (five years afterward would be still stronger evidence of a trait).

Richie's lab used the Trier with seasoned (lifetime average = 9,000 hours) vipassana meditators who did an eight-hour day of meditation and the next day underwent the test.[20] The meditators and their age- and gender-matched comparison group all went through the TSST (as well as a test for inflammation—more on those results in chapter nine, "Mind, Body, and Genome").

Result: the meditators had a smaller rise in cortisol during the stress. Just as important, the meditators *perceived* that dreaded Trier test as less stressful than did the nonmeditators.

This cooled-out, more balanced way of viewing that stressor among the seasoned meditators was not tapped while they were practicing but while they were at rest—our "before." Their ease during both the stressful interview and the formidable mental math challenge seems a genuine trait effect.

Further evidence for this comes from research with these same advanced meditators.[21] The meditators' brains were scanned while they saw disturbing images of people suffering, like burn victims. The seasoned practitioners' brains revealed a lowered level of reactivity in the amygdala; they were more immune to emotional hijacking.

The reason: their brains had stronger operative connectivity between the prefrontal cortex, which manages reactivity, and the amygdala, which triggers such reactions. As neuroscientists know, the stronger this particular link in the brain, the less a person will be hijacked by emotional downs and ups of all sorts.

This connectivity modulates a person's level of emotional reactivity: the stronger the link, the less reactive. Indeed, that relationship is so strong that a person's reactivity level can be predicted by the connectivity. So, when these high-lifetime-hour meditators saw an image of a gruesome-looking burn victim, they had little amygdala reactivity. Age-matched volunteers did not show either the heightened connectivity or the equanimity on viewing the disturbing images.

But when Richie's group repeated this study with people taking the MBSR training (a total of just under thirty hours) plus a bit of daily at-home practice, they failed to find any strengthening of connection between the prefrontal region and the amygdala during the challenge of upsetting images. Nor was there any when the MBSR group simply rested.

While MBSR training did reduce the reactivity of the amygdala, the long-term meditator group showed both this reduced reactivity in the amygdala plus strengthening of the connection between the prefrontal cortex and amygdala. This pattern implies that when the going gets tough—for example, in response to a major life challenge such as losing a job—the ability to manage distress (which depends upon the connectivity between the prefrontal cortex and amygdala) will be greater in long-term meditators compared to those who have only done the MBSR training.

The good news is that this resilience can be learned. What we don't know is how long this effect might last. We suspect that it would be short-lived unless participants continued to practice, a key to transforming a state into a trait.

Among those who show the most short-lived amygdala response, emotions come and go, adaptive and appropriate. Richie's lab put this idea to the test with brain scans of 31 highly seasoned meditators (lifetime average was 8,800 hours of meditation practice, ranging from just 1,200 to more than 30,000).

They saw the usual pictures ranging from people in extreme suffering (burn victims) to cute bunnies. On first analysis of the expert meditators' amygdalae, there was no difference in how they reacted from the responses of matched volunteers who had never meditated. But when Richie's group divided the seasoned meditators into those with the least hours of practice (lifetime average 1,849 hours) and the most (lifetime average 7,118), the results showed that the more hours of practice, the more quickly the amygdala recovered from distress.[22]

This rapid recovery is the hallmark of resilience. In short, equanimity emerges more strongly with extended practice. Among the benefits of long-term meditation, this tells us, are exactly what those Desert Fathers were after: a mind undisturbed.

IN A NUTSHELL

The amygdala, a key node in the brain's stress circuitry, shows dampened activity from a mere thirty or so hours of MBSR practice. Other mindfulness training shows a similar benefit, and there are hints in the research that these changes are traitlike: they appear not simply during the explicit instruction to perceive the stressful stimuli mindfully but even in the "baseline" state, with reductions in amygdala activation as great as 50 percent. Such lessening of the brain's stress reactions appears in response not simply to seeing the gory pictures used in the laboratory but also to more real-life challenges like the stressful Trier interview before a live audience. More daily practice seems associated with lessened stress reactivity. Experienced Zen practitioners can withstand higher levels of pain, and have less reaction to this stressor. A three-month meditation retreat brought indicators of better emotional regulation, and long-term practice was associated with greater functional connectivity between the prefrontal areas that manage emotion and the areas of the amygdala that react to stress, resulting in less reactivity. And an improved ability to regulate attention accompanies some of the beneficial impact of meditation on stress reactivity. Finally, the quickness with which long-term meditators recover from stress underlines how trait effects emerge with continued practice.

6

Primed for Love

In arid landscapes during ancient times, grapes were rare, a succulent delicacy grown in distant regions. Yet one day, records from the second century AD tell us, someone brought just such a treat all the way to the desert abode of Macarius, a Christian hermit.[1]

But Macarius did not eat the grapes; instead he gave them to another hermit nearby who was feeble and who seemed in greater need of the treat.

And that hermit, though grateful for Macarius's kindness, thought of yet another among them who would benefit from eating the grapes, and passed them on to that monk. So it went through the entire hermit community until the grapes came around again to Macarius.

Those early Christian hermits, known as Desert Fathers, lauded the same wholesome modes of being as do yogis in the Himalayas today, who follow surprisingly similar discipline, customs, and meditative practices. They share an ethic of selflessness and generosity and live in isolation, the better to immerse themselves in meditation.

What propelled those juicy grapes' journey through that desert commune? The drivers were compassion and loving-kindness, the attitude of putting the needs of others ahead of our own. Technically, "loving-kindness" refers to wishing that other people be happy; its near cousin "compassion" entails the wish that people be relieved of suffering. Both outlooks (which we'll just refer to as "compassion") can be strengthened through mind training—and if successful, the result will be acting to help others, as demonstrated by the Desert Fathers and that bunch of grapes.

But consider a modern update. Divinity students at a theological seminary were told they would be evaluated on a practice sermon. Half were given a random selection of Bible topics for their sermon. The other half were assigned the parable of the Good Samaritan, the man who stopped to help a stranger in need who was lying by the side of the road, even as others walked by, indifferent.

After a time to prepare their thoughts, they went one by one to another building, where they were evaluated on the talk they had just prepared. As each of them in turn passed through a courtyard on the way to give their sermon, they passed a man who was bent over and moaning in pain.

The question: Did they stop to help the stranger in need?

Turns out whether a divinity student helped or not depended on how late that student felt—the more time-pressured, the less likely to stop.[2] When we are rushing through a busy day, worried about getting to the next place on time, we tend literally not to notice the people around us, let alone their needs.

There's a spectrum that runs from self-centered preoccupations

(*I'm late!*), to noticing the people around us, to tuning in to them, empathizing, and finally, if they are in need, acting to help.

Holding the attitude of compassion means we merely espouse this virtue; *embodying* compassion means we act. The students pondering the Good Samaritan likely were appreciating his compassion—but were not more likely to act with compassion themselves.

Several meditation methods aim to cultivate compassion. The scientific (and ethical) question is, Does this matter—does it move people toward compassionate action?

MAY ALL BEINGS BE FREE FROM SUFFERING

During Dan's first stay in India, in December 1970, he was asked to lecture at a conference on yoga and science in New Delhi. Among the many Western travelers who came to hear Dan was Sharon Salzberg, then an eighteen-year-old doing an independent study year from the State University of New York at Buffalo. Sharon had joined the thousands of young Westerners who made the overland journey from Europe through the Near East to India in the 1970s, travel that warfare and politics have made virtually impossible today.

Dan mentioned that he had just come from a vipassana course given by S. N. Goenka in Bodh Gaya, and that a series of these ten-day retreats was continuing. Sharon was among the handful of Westerners who headed straight from Delhi to the Burmese vihara in Bodh Gaya to take part. She became an ardent student of the method and continued her meditation studies with teachers in India and Burma, and after returning to the States became a teacher herself, cofounding the Insight Meditation Society in Massachusetts—along with Joseph Goldstein, whom she met at the vihara.

Sharon has become the leading advocate of a method she first learned from Goenka, called *metta* in Pali and loosely translated into English as "loving-kindness"—an unconditional benevolence and goodwill—a quality of love akin to the Greek *agape*.[3]

In the format for loving-kindness that Sharon helped bring to the West, you silently repeat phrases like "May I be safe," "May I be healthy," and "May my life unfold with ease," first wishing this for yourself, then for people you love, then for neutral people, and finally for all beings—even those whom you find difficult or who have harmed you. In one version or another, this has become the most well-studied format of compassion meditation.

This version of loving-kindness sometimes includes the compassionate wish that people be free from suffering, too. And though the difference between loving-kindness and compassion may be consequential in some way, little attention gets paid to this distinction in the research world.

Years after her return from India, Sharon was a panelist in a dialogue with the Dalai Lama in 1989, for which Dan was moderator.[4] At one point Sharon told the Dalai Lama that many Westerners felt loathing toward themselves. He was astonished—he'd never heard of this. He had, the Dalai Lama replied, always assumed that people naturally loved themselves.

Yet in English the word *compassion*, the Dalai Lama pointed out, signifies the wish that others be well—but it does not include oneself. He explained that in his own language, Tibetan, as well as in the classical tongues Pali and Sanskrit, the word *compassion* implies feeling this for oneself as well as others. English, he added, needs a new word, *self-compassion.*

That very term came into the world of psychology more than a decade later when Kristin Neff, a psychologist at the University of Texas at Austin, published her research on a measure of self-compassion. In her definition this includes being kind to yourself instead of self-critical; seeing your failures and mistakes as just part of the human condition rather than some personal failing; and just noting your imperfections, not ruminating about them.

The opposite of self-compassion can be seen in the constant self-criticism common, for example, in depressed ways of thinking. Loving-kindness directed to yourself, on the other hand, would seem to offer a direct antidote. An Israeli group tested this idea, and found that teaching loving-kindness to people particularly prone to self-criticism both lessened those harsh thoughts and increased their self-compassion.[5]

EMPATHY MEANS FEELING WITH

Brain research tells us of three kinds of empathy.[6] Cognitive empathy lets us understand how the other person thinks; we see their perspective. In emotional empathy we feel what the other is feeling. And the third, empathic concern or caring, lies at the heart of compassion.

The word *empathy* entered the English language only in the early years of the twentieth century, as a translation of the German word *Einfühlung,* which might be translated as "feeling with." Purely cognitive empathy has no such sympathetic feelings, while the defining sign of emotional empathy is feeling in your own body what the suffering person seems to feel.

But if what we feel upsets us, all too often our next response means we tune out, which helps us feel better but blocks compassionate action. In the lab one way this withdrawal instinct shows up is in people averting their gaze from photos that depict intense suffering—like a man so painfully burned that his skin has peeled away. Similarly, homeless people complain that they become invisible—those passing by on the street ignore them, another form of averting the gaze from suffering.

Since compassion begins with accepting what's happening without turning away—an essential first step toward taking helpful action—could meditations that cultivate compassion tip the balance?

Researchers at Germany's Max Planck Institute in Leipzig taught volunteers a version of loving-kindness meditation.[7] The volunteers practiced generating such loving-kindness in a six-hour instructional session, and at home on their own.

Before they had learned this loving-kindness method, when the volunteers saw graphic videos of people suffering, only their negative circuits for emotional empathy activated: their brains reflected the state of the victims' suffering as though it were happening to themselves. This left them feeling upset, an emotional echo of distress that transferred from the victims to themselves.

Then people were instructed to empathize with the videos—to share the emotions of the people they were seeing. Such empathy, fMRI studies revealed, activated circuits centering on parts of the insula—circuits that light up when we ourselves are suffering. Empathy meant that people felt the pain of those who were suffering.

But when another group instead got instructions in compassion—

feeling love for those suffering—their brains activated a completely different set of circuits, those for parental love of a child.[8] Their brain signature was clearly different from those who received instructions in empathy.

And this after only eight hours!

Such positive regard for a victim of suffering means we can confront and deal with their difficulty. This allows us to move along that spectrum from noticing what's going on to the payoff, actually helping them. In many East Asian countries the name Kuan Yin, the revered symbol of compassionate awakening, translates as "the one who listens and hears the cries of the world in order to come and help."[9]

FROM ATTITUDE TO ACTION

The skeptical scientist has to ask, Does displaying this neural pattern mean people will actually help, especially if doing so means they have to do something uncomfortable, even make a sacrifice? Just measuring brain activity in people while they lie still in a brain scanner, and even finding that neural priming for kindness and action gets stronger, is intriguing but not convincing. After all, those seminary students reflecting on the Good Samaritan were not more likely to actually help someone in need.

But some evidence suggests a more hopeful outcome. In Richie's lab, volunteers' brains were scanned before and then after two weeks of either compassion training (thinking of others) or cognitive reappraisal, a self-focus, in which you are taught to think differently about the causes of negative events. Then their brains were scanned as they viewed images of human suffering. After the brain scan they played the Redistribution Game, where they first witnessed a "dictator" cheat a victim out of a fair share of $10, giving just one measly dollar. The game then let the volunteers give up to $5 of their own money to the victim, and game rules forced the dictator to give twice that amount to the victim.

Result: those with the training in compassion gave almost two times as much to the victim as did the group who had learned how to reappraise their feelings. And their brain showed increased activation in circuits for attention, perspective taking, and positive feelings; the more of this activation, the more altruistic.

As Martin Luther King Jr. commented on the Good Samaritan tale, those who did not help asked themselves, If I stop to help, what will happen to me?

But the Good Samaritan asked, If I don't stop to help, what will happen to him?

READY TO LOVE

Anyone with half a heart would find it painful to look at a photo of a young child on the brink of starvation, his large, sad eyes downcast, his mood sullen, his stomach distended while his bones show through his emaciated body.

That image, like the one of the burn victim, has been used in several of these studies of compassion as part of a standard test of the ability to confront suffering. In the arc from ignoring someone's pain or need, to noticing, empathizing, and then acting to help, stirring up feelings of loving-kindness energizes every step.

Studies with novices learning loving-kindness reveal an early harbinger of heightened amygdala reactions to images of pain and suffering found in seasoned meditators.[10] The finding was nowhere as strong as in the long-term meditators—just a hint that the pattern can show up very soon.

How soon? Maybe in mere minutes—at least when it comes to mood. One study found that just seven minutes of loving-kindness practice boosts a person's good feelings and sense of social connection, if only temporarily.[11] And the Davidson group had found that after eight or so hours of training in loving-kindness, volunteers showed strong echoes of those brain patterns found in more experienced meditators.[12] The beginners' temporary wave of mellow feeling may be an early precursor of the more striking brain changes in people who practice loving-kindness for weeks, months, or years.

And consider a random group of people who volunteered to try web-based instruction in meditation, for a total of two and a half hours (that is, twenty sessions of ten minutes each). This brief loving-kindness training resulted in people feeling more relaxed and donating to charity at a higher rate than those in a comparison group who did a comparable amount of light exercise like stretching.[13]

Pulling together findings from Richie's lab among others, we can piece together a neural profile of reactions to suffering. Distress circuitry connecting to the insula, including the amygdala, responds with particular strength—a pattern typical of anyone's empathy with other people's pain. The insula monitors the signals in our body and also activates autonomic responses like heart rate and breathing—as we empathize, our neural centers for pain and distress echo what we pick up from the other person. And the amygdala signals something salient in the environment, in this case, the suffering of another. The more deeply immersed in the compassion meditation a person reported being, the stronger was this empathic pattern—compassion seems to amplify empathy to suffering, just as that meditation intends.

In a different study from Richie's lab, long-term meditators generating compassion showed a strong increase in the amygdala response to distressing sounds (like a woman's scream), while for those in a comparison group there was little difference between compassion and the neutral control condition.[14] In a companion study, participants had brain scans while concentrating on a small light as they heard those disturbing sounds.[15] In meditation-naive volunteers, the amygdala flared into action at those sounds, while in the meditators the amygdala response was muted and their concentration strong. Even those volunteers who had been promised a reward if they exerted effort in focusing on the light no matter what they heard nonetheless were distracted by the screams.

Putting these findings together gives several clues about how mental training works. For one thing, very often meditation comes in batches, not as a single practice. Vipassana meditators (the majority of those in the long-term studies reported here) on a typical retreat might mix mindfulness of breathing with loving-kindness. MBSR and similar programs offer several kinds of mental training.

These various mind training methods drive the brain in different ways. During compassion practice, the amygdala is turned up in volume, while in focused attention on something like the breath, the amygdala is turned down. Meditators are learning how to change their relationship to their emotions with different practices.

The amygdala's circuits light up when we are exposed to someone feeling a strong negative emotion—fear, anger, and the like. This amygdala signal alerts the brain that something important is happening; the amygdala acts as neural radar detecting the salience of whatever we experience. If what's going on seems urgent, like a woman screaming in fear, the amygdala has extensive connections to recruit other circuitry to respond.

Meanwhile the insula uses its connections to the body's visceral organs (like the heart) to ready the body for active engagement (increasing blood flow to the muscles, for example). Once the brain primes the body to respond, those who have meditated on compassion are more likely to act to help someone.

But then there's the question of how long such effects of mental training in compassion last. Is this only a temporary state, or does it become a lasting trait? Seven years after his three-month retreat experiment ended, Cliff Saron tracked down the participants.[16] He found a surprise among those who, during and just after the retreat, were able to sustain attention to disturbing images of suffering—a psychophysiological measure of acceptance, as opposed to the averted gaze and expression of disgust he found in others (and which typifies people in general).

Those who did not avert their eyes but took in that suffering were, seven years later, better able to remember those specific pictures. In cognitive science, such memory betokens a brain that was able to resist an emotional hijack, and so, take in that tragic image more fully, remember it more effectively—and, presumably, act.

Unlike other benefits of meditation that emerge gradually—like a quicker recovery from stress—enhancing compassion comes more readily. We suspect that cultivating compassion may take advantage of "biological preparedness," a programmed readiness to learn a given skill, as seen, for instance, in the rapidity with which toddlers learn language. Just as with speaking, the brain seems primed to learn to love.

This seems largely due to the brain's caretaking circuitry, which we share with all other mammals. These are the networks that light up when we love our children, our friends—anyone who falls within our natural circle of caring. These circuits, among others, grow stronger even with short periods of compassion training.

As we've seen, enhancing a compassionate attitude goes beyond a mere outlook; people actually grow more likely to help someone in need even when there's a cost to themselves. Such intense resonance with others' suffering has been found in another notable group: extraordinary altruists, people who donated one of their kidneys to a stranger in dire need of a transplant. Brain scans discovered that these compassionate souls have a larger right-side amygdala compared to other people of their age and gender.[17]

Since this region activates when we empathize with someone who is suffering, a larger amygdala may confer an unusual ability to feel the pain of others, so motivating people's altruism—even as extraordinarily as donating a kidney to save someone's life. The neural changes from loving-kindness practice (the emerging signs of which are found even among beginners) align with those found in the brains of the super-Samaritan kidney donors.[18]

The cultivation of a loving concern for other people's well-being has a surprising and unique benefit: the brain's circuitry for happiness energizes, along with compassion.[19] Loving-kindness also boosts the connections between the brain's circuits for joy and happiness and the prefrontal cortex, a zone critical for guiding behavior.[20] And the greater the increase in the connection between these regions, the more altruistic a person becomes following compassion meditation training.

NURTURING COMPASSION

When she was young, Tania Singer thought she might have a career on the stage, perhaps as director of theater and opera. And from her college years on, she plunged into meditation retreats of different sorts, studying with a variety of teachers as the years went on. The methods ranged from vipassana to Father David Stendl-Rast's practice of gratitude. She was drawn to teachers who embodied a quality of unconditional love.

The mysteries of the human mind drew Tania into psychology, the field in which she earned her PhD; her doctoral research on learning in very old age got her interested in plasticity research. Her postdoc research on empathy revealed that when we witness the pain and suffering of someone else we activate networks which underlie these very same feelings in ourselves—a discovery that got wide attention, laying the groundwork for empathy research in neuroscience.[21]

Our empathic resonance with the pain of others, she found, activates what amounts to a neural alarm that instantly tunes us to others' suffering, potentially alerting us to the presence of danger. But compassion—feeling *concern* for the person suffering—seemed to involve a different set of brain circuits, those for feelings of warmth, love, and concern.

This discovery originated from experiments Tania did with Matthieu Ricard, a Tibetan monk with a PhD in science—and decades of meditation practice. Tania asked him to try a variety of meditative states while in a brain scanner. She wanted to see what happened in the brain of an expert meditator in order to design meditation practices anyone could try.

When he cultivated empathy, sharing the suffering of another, she saw the action in his neural networks for pain. But once he began to generate compassion—loving feelings for someone who was suffering—he activated brain circuitry for positive feelings, reward, and affiliation.

Tania's group then reverse engineered what they found with Matthieu by training groups of meditation first-timers in empathizing with a person's suffering, or feeling compassion for their suffering.

Compassion, she found, muted the empathic distress that can lead to emotional exhaustion and burnout (as happens sometimes in the caring professions like nursing). Instead of simply feeling with the other person's angst, compassion training led to that activation of completely different brain circuits, those for loving concern—and to positive feelings and resilience.[22]

Now Tania directs the Department of Social Neuroscience at the Max Planck Institute for Human Cognitive and Brain Sciences, in Leipzig, Germany. In a melding of her meditative and scientific interests and based on her previous promising plasticity research on empathy and compassion training, Tania has done definitive research on meditation as a way to cultivate wholesome mental qualities such as attention, mindfulness, perspective taking, empathy, and compassion.

In an elegant program of research called the ReSource Project Tania's group recruited around three hundred volunteers who committed to spending eleven months in different types of contemplative practices, practicing each in three modules of several months—plus a comparison group that got no training but took the same battery of tests every three months.

The first mental training, "Presence," entailed a body scan and breath focus. Another, "Perspective," included observing thoughts via a novel interpersonal practice of "contemplative dyads," where partners share their stream of thought with each other for ten minutes daily, either through a cell phone app or in person.[23] The third, "Affect," included loving-kindness practice.

Results: the scan increased body awareness and lessened mind-wandering. Observing thoughts enhanced meta-awareness, a by-product of mindfulness. On the other hand, loving-kindness boosted warm thoughts and feelings about others. In short, if you want to increase your feelings of kindness most effectively, practice exactly that—not something else.

WHAT'S THE ACTIVE INGREDIENT?

"Samantha has HIV," you read. "She contracted the disease from a dirty needle in a doctor's office abroad. She attends peace rallies once a month. She did well in high school." Next to this thumbnail sketch you see Samantha's photo, revealing a twenty-something woman with shoulder-length hair.

Would you donate money to help her out?

To learn what inner factors are at work here, researchers at the University of Colorado taught a compassion meditation to one set of volunteers, while an ingenious control group took a daily puff of "placebo oxytocin," a phony feel-good brain drug, which they had been told would increase their feelings of connection and compassion. The phony drug created positive expectations matching that of the compassion meditators.[24]

After either the meditation or the puff, a mobile phone app showed each person a picture and thumbnail profile of someone in need like Samantha, with the option to donate to them some of the money the volunteer was being paid.

Tellingly, simply doing the compassion meditation was not the strongest predictor of whether someone donated. In fact, in this study those doing the compassion meditation were no more likely to donate than those who puffed the fake oxytocin—or a group who did neither. Not to get too geeky, but this raises a key point about the methods used in meditation research. While this study had a first-rate design in many respects (such as that clever fake oxytocin control group), in at least one way the study is murky: the nature of the compassion meditation was unspecified, seems to have changed over the course of the study, and included meditation that cultivates equanimity.

These contemplative exercises were taken from a set designed to help people working with the dying (pastoral counselors, hospice workers) stay sensitive to suffering while feeling equanimity toward a dying person—after all, there is little or no help to give at that point, save a compassionate presence. And while they were no more likely to donate money, those who did the compassion meditation felt more tenderness toward the people in need. We wonder whether equanimity may have a very different effect on donations than does compassion—perhaps making someone less likely to, say, give money, even while resonating with the suffering.

This raises a related issue, whether you need to focus on loving-kindness to enhance compassionate acts. For example, at Northeastern University, volunteers were taught either mindfulness or loving-kindness meditation.[25] After two weeks of lessons, each found themselves in a waiting room with a woman on crutches and in apparent pain; two other people on chairs ignored her, and there were only three chairs. As in the Good Samaritan study, each of the meditators had the choice to give their own chair so the person on crutches could sit down.

Both those who had learned mindfulness and those who practiced loving-kindness—compared to a group who did neither—more often took the route of kindness, giving up their chair (in the nonmeditating control group, 15 percent gave up their chair, while for the meditators it was around 50 percent). But from this study alone we don't know whether mindfulness enhances empathy just like loving-kindness practice, or if other inner forces—like a greater attention to circumstances—compelled that act of compassion.

First signs suggest that each variety of meditation has its own neural profile. Take results from research spearheaded by Geshe Lobsang Tenzin Negi, who has a degree in the philosophical and practice tradition shared by the Dalai Lama (a Tibetan *geshe* is the equivalent of our PhD), as well as a PhD from Emory University, where he teaches. Geshe Negi drew on his background as a scholar and monk to create Cognitively-Based Compassion Training (CBCT), methods for understanding how one's attitudes support or hinder a compassionate response. This includes a variety of loving-kindness meditation, aspiring to help others be happy and free from suffering, and the determination to act accordingly.[26]

In research at Emory, one group did CBCT, while the other did Alan Wallace's method of meditation (we described this in chapter five, "A Mind Undisturbed"). The main finding: the compassion group's right amygdala tended to increase its activity in response to photos of suffering, and the more hours of practice, the larger the response. They were sharing the suffering person's distress.

But on a test of depressive thinking, the compassion group also reported being happier in general. Sharing another person's feelings of distress need not be a downer. As Dr. Aaron Beck, who designed that depression test, has said, when you focus on someone else's suffering, you forget your own troubles.

Then there's the gender difference. The Emory University researchers, for example, found women show higher levels of right amygdala reactivity than do men in response to all emotional images, happy or sad, including those of suffering. This finding is not exactly news in psychology; brain studies have long shown women are more attuned to other people's emotions than are men.[27] This may be another case of science proving the obvious: women, on average, seem to be more responsive to other people's emotions than men.[28]

Paradoxically, women do not seem more likely than men actually to act when confronted with an opportunity to help, perhaps because they sometimes feel more vulnerable.[29] There are more factors at work in compassionate action than simply a brain signature, a fact that researchers in this area continue to struggle with. Factors from feeling pressured for time, to whether you identify with the person in need, to whether you are in a crowd or alone—each of these factors can matter. One open question: Will cultivating a compassionate outlook prime a person sufficiently to overcome these other forces in the face of someone's need?

WIDENING OUR CIRCLE OF CARING

A highly accomplished Tibetan meditation master studied in Richie's lab once said that one hour spent practicing loving-kindness toward a difficult person is equivalent to one hundred hours of the same toward a friend or loved one.

The generic loving-kindness meditation takes us through an ever-widening circle of the kinds of people we try to hold tender feelings toward. The biggest leap comes when we extend love beyond people we know and love, to people we don't know, let alone those we find difficult. And then after that there's the grand aspiration to love everyone, everywhere.

How can we extend the compassion we feel for our immediate loved ones to the entire human family, including people we don't like? This big leap in loving-kindness—were it to become more than a mere wish—might go far in healing many divides in the world that cause pain and conflict.

The Dalai Lama sees one strategy: recognize the "oneness" of humankind, even groups we dislike, and so realize that "all of them, like ourselves, do not want suffering; they want happiness."[30]

Does this feeling of oneness help? We don't know yet, from a research viewpoint. Easy to say but hard to do. One strict test of this shift toward universal love might measure unconscious bias—when you act outside your awareness in a prejudiced way toward some group, despite believing you harbor no such animus.

These hidden biases can be detected via clever tests. For example, a person may say he has no racial prejudice, yet when presented with a reaction time test in which words that have pleasant or unpleasant connotations are paired with the words *black* or *white*, words with pleasant meaning are more quickly paired with the word *white* compared with the word *black*, and vice versa.[31]

Researchers at Yale University used such a measure of implicit bias before and after a six-week class in loving-kindness meditation.[32] This research used a strong control group—teaching participants about the value of loving-kindness meditation without actually teaching them the practice. A bit like those divinity students pondering the Good Samaritan, this no-practice group showed zero benefit on the implicit bias test. The drop in unconscious prejudice came from loving-kindness.

The Dalai Lama tells of his half century of working at cultivating compassion. At the start, he says, he had enormous admiration for those who had developed genuine compassion for all beings—but he was not confident he could do so himself.

He knew intellectually that such unconditional love was possible, but that it took a certain kind of inner work to build up. As time went on, he found that the more he practiced and became familiar with the feelings of compassion, the stronger his courage became that he, too, could develop it at the higher levels.

With this penultimate variety of compassion, he adds, we are impartial in our concern, extending it toward everyone, everywhere—even when those we feel it toward hold animosity toward us. What's more, ideally this feeling does not come just sporadically, from time to time, but has become a compelling and stable force, a central organizing principle of our lives.

And whether or not we attain that lofty height of love, there are other benefits along the way, like how the brain's circuitry for happiness energizes, along with compassion. As we've often heard the Dalai Lama say, "The first person to benefit from compassion is the one who feels it."

The Dalai Lama recalls an encounter at Montserrat, a monastery near Barcelona, with Padre Basili, a Christian monk who had been in isolated retreat in a nearby mountain hermitage for five years. What had he been doing?

Meditating on love.

"I noticed a glow in his eyes," the Dalai Lama said, adding this indicated the depth of his peace of mind and the beauty from becoming a wonderful person. The Dalai Lama noted that he had met people who had everything they wanted, yet were miserable. The ultimate source of peace, he said, is in the mind—which, far more than our circumstances, determines our happiness.[33]

IN A NUTSHELL

Simply learning about compassion does not necessarily increase compassionate behavior. In the arc from empathizing with someone suffering to actually reaching out to help, loving-kindness/compassion meditation ups the odds of helping. There are three forms of empathy—cognitive empathy, emotional empathy, and empathic concern. Often people empathize emotionally with someone's suffering but then tune out to soothe their own uncomfortable feelings. But compassion meditation enhances empathic concern, activates circuits for good feelings and love, as well as circuits that register the suffering of others, and prepares a person to act when suffering is encountered. Compassion and loving-kindness increase amygdala activation to suffering while focused attention on something neutral like the breath lessens amygdala activity. Loving-kindness acts quickly, in as little as eight hours of practice; reductions in usually intractable unconscious bias emerge after just sixteen hours. And the longer people practice, the stronger these brain and behavioral tendencies toward compassion become. The strength of these effects from the early days of meditation may signal our biological preparedness for goodness.

Attention!

One day a student asked his Zen teacher to create a brushstroke calligraphy for him, "something of great wisdom."

The Zen master, without hesitating, took up his brush and wrote: *Attention.*

His student, a bit dismayed, asked, "Is that all?"

Without a word, the master took to his brush again, and wrote, *Attention. Attention.*

His student, feeling that was not so profound, got a bit irritated, complaining to the master there was nothing so wise about that.

Again the master responded in silence, writing *Attention. Attention. Attention.*

Frustrated, the student demanded to know what he meant by that word, *attention*. To which the master replied, "Attention means attention."[1]

William James made explicit what that Zen master may have been hinting at: "The faculty of bringing back a wandering attention over and over again is the very root of judgment, character and will," he declared in his *Principles of Psychology*, published in 1890. James went on to say that "an education which should improve this faculty would be *the* education *par excellence*."

After that bold claim he backtracked a bit adding, "But it is easier to define this ideal than to give practical directions for bringing it about."

Richie had read this passage before he went to India, and after his transformative moments at the Goenka retreat, those words flashed back in his mind with an electric charge.

This was a seminal moment, an intellectual pivot point for Richie. He had the gut sense that he had found that most excellent education James sought: meditation. Whatever specific form it takes, most every kind of meditation entails retraining attention.

But the research world knew little about attention back in our graduate school days in the 1970s. The one study that connected meditation to an improvement in attention was by Japanese researchers.[2] They brought an EEG machine to a zendo and measured monks' brain activity during meditation while hearing a monotonous series of sounds. While most monks showed nothing remarkable, three of the most "advanced" monks did: their brains responded as strongly to the twentieth sound as to the first. This was big news: ordinarily the brain would tune out, showing no reaction to the tenth *bing*, let alone the twentieth.

Tuning out a repeated sound reflects the neural process known as habituation. Such waning in attention to anything monotonous can plague radar operators, who have to stay vigilant while scanning signals from mostly empty sky. Attention fatigue in radar operators was the practical reason this very aspect of attention had been intensively researched during World War II, when psychologists were asked how to keep operators alert. Only then did attention come under scientific study.

Ordinarily we notice something unusual just long enough to be sure it poses no threat, or simply to categorize it. Then habituation conserves brain energy by paying no attention to that thing once we know it's safe or familiar. One downside of this brain dynamic: we habituate to *anything* familiar—the pictures on our walls, the same dish night after night, even, perhaps, our loved ones. Habituation makes life manageable but a bit dull.

The brain habituates using circuitry we share even with reptiles: the brain stem's reticular activating system (RAS), one of the few attention-related circuits known at the time. In habituation, cortical circuits inhibit the RAS, keeping this region quiet when we see the same old thing over and over.

In the reverse, sensitization, as we encounter something new or surprising, cortical circuits activate the RAS, which then engages other brain circuits to process the novel object—a new piece of art in place of a too-familiar one, say.

Elena Antonova, a British neuroscientist who has attended the SRI, found that meditators who had done a three-year retreat in the Tibetan tradition had less habituation of eye blinks when they heard loud bursts of noise.[3] In other words, their blinks continued unabated. This replicates (at least conceptually) that study from Japan where advanced Zen meditators did not habituate to repetitive sounds.

The original Zen study was for us seminal. It seemed the Zen brains could sustain attention when other brains would tune out. This resonated with our own experience at retreats on mindfulness, where we spent hours pushing our attention to notice every little detail of experience rather than tune out.

By zooming in on details of sights, sounds, tastes, and sensations that we otherwise would habituate to, our mindfulness transformed the familiar and habitual into the fresh and intriguing. This attention training, we saw, might well enrich our lives, giving us the choice to reverse habituation by focusing us on a deeply textured here and now, making "the old new again."

Our early view of habituation saw mindfulness as a voluntary shift from the reflexive tune out. But that was as far as we had gotten in our thinking—and was already pushing the boundaries of accepted scientific thought. Back in the 1970s science saw attention as mostly stimulus-driven, automatic, unconscious, and from the "bottom up"— a function of the brain stem, a primitive structure sitting just above the spinal cord, rather than from a "top-down" cortical area.

This view renders attention involuntary. Something happens around us—a phone rings—and our attention automatically gets pulled to the source of that sound. A sound continues to the point of monotony and we habituate.

There was no scientific concept for the volitional control of attention—despite the fact that psychologists themselves were using their volitional attention to write about how no such ability existed! In keeping with the scientific standards of the day, the reality of their own experience was simply ignored in favor of what could be objectively observed.

This truncated view of attention gave only part of the story. Habituation describes one variety of attention over which we have no conscious control, but higher in our neural circuitry, above these bottom-of-the-brain mechanisms, different dynamics apply.

Take the emotional centers in the midbrain's limbic system, where much of the action originates when emotions drive our attention. When Dan wrote *Emotional Intelligence,* he drew heavily on research by Richie and other neuroscientists on the then new discovery of the dance of the amygdala, the brain's radar for threat (in the midbrain's emotion circuits) with prefrontal circuitry (behind the forehead) the brain's executive center, which can learn, reflect, decide, and pursue long-term goals.

When anger or anxiety is triggered, the amygdala drives prefrontal circuitry; as such disturbing emotions reach their peak, an amygdala hijack paralyzes executive function. But when we take active control of our attention—as when we meditate—we deploy this prefrontal circuitry, and the amygdala quiets. Richie and his team have found this quieted amygdala both in long-term vipassana meditators and, with hints of the same pattern—though less strong—in people after training in MBSR.[4]

Richie's scientific career has tracked the locus of attention as it moved steadily up the brain. In the 1980s he helped found affective neuroscience, the field that studies the midbrain's emotional circuitry and how emotions push and pull attention. By the 1990s, as contemplative neuroscience began and researchers started looking at the brain during meditation, they knew how circuitry in the prefrontal cortex manages our voluntary attention. This area has today become the brain's hot spot for meditation research; every aspect of attention involves the prefrontal cortex in some way.

In humans the prefrontal cortex takes up a larger ratio of the brain's top layer, the neocortex, than in any other species, and has been the site of the major evolutionary changes that make us human. This neural zone, as we will see, holds the seeds of awakening to enduring well-being, but it is also entwined with emotional suffering. We can envision wonderful possibilities, and we also can be disturbed by worrisome thoughts—both signs of the prefrontal cortex at work.

While William James wrote about attention as though it were one single entity, science now tells us the concept refers not just to one ability but to many. Among them:

> *Selective* attention, the capacity to focus on one element and ignore others.
>
> *Vigilance,* maintaining a constant level of attention as time goes on.
>
> *Allocating* attention so we notice small or rapid shifts in what we experience.
>
> *Goal focus,* or "cognitive control," keeping a specific goal or task in mind despite distractions.
>
> *Meta-awareness*, being able to track the quality of one's own awareness—for example, noticing when your mind wanders or you've made a mistake.

SELECTING ATTENTION

From infancy, Amishi Jha can remember her parents meditating every morning using beads to recite mantra, as they had learned in their native India. But Amishi was not interested in meditation; she went on to become a cognitive neuroscientist trained in the rigorous study of attention.

While Amishi was on faculty at the University of Pennsylvania, Richie came to lecture. During his talk he never mentioned meditation, but he did show fMRI images of two brains—one in the depths of depression, the other happy. Amishi asked him, "How do you get a brain to change from one to the other?"

"Meditation," Richie answered.

That got Amishi's interest, both personally and professionally. She started to meditate, and began to do research on how the method might impact attention. But she got pushback from her colleagues, who cautioned her that it was too risky and might not be of broad scientific interest within the field of psychology.

The next year she attended the second meeting of the Mind and Life Summer Research Institute, which proved transformative. The faculty, graduate students, and postdocs she met there were a supportive community, who encouraged Amishi.

Richie vividly remembers an emotional testimonial Amishi gave at this meeting about how meditation was part of her root culture. While she had felt constrained in pursuing such research in the academy, she felt she finally found her home with like-minded scientists doing research in this area. Amishi has become a leader of a new generation of scientists committed to contemplative neuroscience and its benefits for society.

She and her colleagues conducted one of the first rigorous studies of how meditation impacts attention.[5] Her lab, now based at the University of Miami, found that novices trained in MBSR significantly improved in orienting, a component of selective attention that directs the mind to target one among the virtually infinite array of sensory inputs.

Let's say you are at a party listening to the music, and tuning out a conversation going on right next to you. If someone were to ask you what they had just said, you'd have no idea. But should one of them mention your name, you would zero in on those dulcet sounds as though you had been listening to them right along.

Known in cognitive science as the "cocktail party effect," this sudden awareness illustrates part of the design of our brain's attention systems: we take in more of the stream of information available than we know in conscious awareness. This lets us tune out irrelevant sounds but still examine them for relevance somewhere in the mind. And our own name is always relevant.

Attention, then, has various channels—the one we consciously select and those we tune out of. Richie's dissertation research examined how meditation might strengthen our ability to focus as we choose by asking volunteers to pay attention to what they saw (a flashing light) and ignore what they felt (a vibration on the wrist) or vice versa, while he used EEG readings of their visual or tactile cortex to measure the strength of their focus. (His use of EEG to examine this in humans, by the way, broke new ground—it had only been done with rats and cats until then.)

The meditators among the volunteers showed a modest increase in what he called "cortical specificity"—more activity in the appropriate part of the sensory areas of the cortex. So, for example, when they were paying attention to what they saw, the visual cortex was more active than the tactile.

When we choose to concentrate on visual sensations and ignore what we touch, the lights are "signal" and the touch "noise." When we get distracted, noise drowns the signal; concentration means much more signal than noise. Richie found no increase in the signal, but there was some reduction in noise—altering the ratio. Less noise means more signal.

Richie's dissertation study, like Dan's, was slightly suggestive of the effect he was seeking to find. Fast-forward several decades to far more sophisticated measures of the well-targeted sensory awareness Richie had tried to demonstrate. A group at MIT deployed MEG—a magnetic EEG measure with a much more precise targeting of brain areas than Richie's old-time EEG had allowed—with volunteers who had been randomly assigned to either get an eight-week program in MBSR, or who waited to get the training until after the experiment was done.[6]

MBSR, remember, includes mindfulness of breath, practicing a systematic scan of sensations throughout your body, attentive yoga, and moment-to-moment awareness of thoughts and feelings—with the invitation to practice these attention training methods daily. After eight weeks those who had gone through the MBSR program showed a far better ability to focus on sensations—in this case a carefully calibrated tapping on their hand or foot—than they had done before starting the MBSR training, as well as better than those who were still waiting for MBSR.

Conclusion: mindfulness (at least in this form) strengthens the brain's ability to focus on one thing and ignore distractions. The neural circuitry for selective attention, the study concluded, can be trained—contrary to the standard wisdom where attention was assumed to be hardwired and so, beyond the reach of any training attempt.

A similar strengthening of selective attention was found in vipassana meditators at the Insight Meditation Society who were tested before and after a three-month retreat.[7] The retreat offered what amounts to explicit encouragement to be fully attentive, not just in the daily eight hours of formal sittings but throughout the day as well.

Before the retreat, when they paid attention to selective beeps or boops, each at a different tone, their accuracy in spotting the target tones was no better than anyone else's. But after three months the retreatants' selective attention was markedly more accurate, showing more than a 20 percent gain.

SUSTAINING ATTENTION

Zen scholar D. T. Suzuki was a panelist at a symposium held outdoors. As he sat behind a table with the other panelists, Suzuki was perfectly still, his eyes fixed on a spot somewhere in front of him, seemingly zoned out in some world of his own. But when a sudden gust of wind blew some papers across the table, Suzuki alone among the panelists made a lightning grab for them. He wasn't zoned out—he was paying keen attention in the Zen fashion.

The ability to sustain attention without habituating in Zen meditators, remember, was one of the meager scientific findings about meditation back when we began this scientific quest. That Zen study, though it had its limitations, spurred us on.

Attention flows through a meager bottleneck in the mind, and we allot that narrow bandwidth parsimoniously. The lion's share goes to what we choose to focus on in the moment. But as we keep our attention on that thing, our focus inevitably wanes, our mind wandering off to other thoughts and the like. Meditation defies this mental inertia.

A universal goal in meditation of every kind comes down to sustaining attention in a chosen way or to a given target, such as the breath. Numerous reports, both anecdotal and scientific, support the idea that meditation leads to better sustained attention, or, to use the technical term, vigilance.

But a skeptic might ask, Is it the meditation practice that enhances attention, or some other factor? That, of course, is why control groups are needed. And to show even more convincingly that the link between meditation and sustained attention is not mere association, but rather a causal one, requires a longitudinal study.

That higher bar was met by Clifford Saron and Alan Wallace's study, where volunteers attended a three-month meditation retreat with Wallace as teacher.[8] They practiced focusing on their breath five hours per day and Saron tested them at the beginning of the retreat, one month into it, at the end, and finally five months later.

The meditators improved in vigilance, with the greatest gains in the first month of retreat. Five months after the retreat ended, each meditator took a follow-up test of vigilance, and, impressively, the improvement gained during retreat was still strong.

To be sure, the gain was likely maintained by the hour of practice daily these meditators reported. Still, this is among the best direct tests of a meditation-induced altered trait in attention we have so far. Of course the evidence would be all the more compelling if these meditators were to show the same gain five years later, too!

WHEN ATTENTION BLINKS

Watch a four-year-old intently scan the crowd in a *Where's Waldo?* drawing, and see the moment of joy when she finally picks out Waldo in his distinctive red-and-white-striped sweater from the confusing crowd. That happy excitement over spotting Waldo marks a key moment in the workings of attention; the brain rewards us for any such victory with a dose of pleasing neurochemicals.

For those few moments, research tells us, the nervous system takes our focus off-line and relaxes, in what amounts to a short neural celebration party. If another Waldo were to pop up during the party, our attention would be occupied elsewhere. That second Waldo would go unseen.

This moment of temporary blindness is like a blink in attention, a short pause in our mind's ability to scan our surroundings (technically, a "refractory period"). During that blink, the mind's ability to notice goes blind and attention loses sensitivity. A slight change that might otherwise catch our eye goes by unnoticed. The blink measure reflects "brain efficiency," in the sense that not getting too caught up in one thing leaves our finite attentional resources available for the next.

Speaking practically, the lack of blink reflects a greater ability to notice small changes—e.g., nonverbal emotional cues of a person's shifting emotions telegraphed by fleeting shifts in the small muscles around the eyes. Insensitivity to such minor signals can mean we miss major messages.

In one test of the blink, you see a long string of letters interspersed with occasional numbers. Each individual letter or number is presented very briefly—for 50 milliseconds, which is 1/20 of a second, at a breathless rate of ten per second. You are warned that each string of letters will contain one or two numbers, at random intervals.

After each string or fifteen or so, you are asked if you saw any numbers and what they were. If two numbers were presented in a rapid-fire sequence most people tend to miss the second number. That's the attentional blink.

Scientists who study attention had long thought this gap in attention immediately after spotting a long-sought target was hardwired, an aspect of the central nervous system that was inevitable and unchangeable. But then something surprising happened.

Enter the meditators at the annual three-month vipassana course at the Insight Meditation Society, the same ones who did so well on the test of selective attention. Vipassana meditation, on the face of it, might lessen the blink, since it cultivates a continuous nonreactive awareness of whatever arises in experience, an "open-monitoring" receptive to all that occurs in the mind. An intensive vipassana course creates something akin to mindfulness on steroids: a nonreactive hyperalertness to all the stuff that arises in one's mind.

Richie's group measured the attentional blink in vipassana meditators before and after that three-month retreat. After the retreat there was a dramatic reduction, 20 percent, in the attentional blink.[9]

The key neural shift was a drop in response to the first glimpse of a number (they were just noting its presence) so the mind remains calm enough to also notice the second number, even if very soon after the first one.

That result was a huge surprise to cognitive scientists, who had believed the attentional blink was hardwired and so could not be lessened by any kind of training. Once the news was out in science circles, a group of researchers in Germany asked whether meditation training might offset the universal worsening with age of the attentional blink, which becomes more frequent and creates longer gaps in awareness as people get older.[10] Yes: meditators who regularly practiced some form of "open monitoring" (a spacious awareness of whatever comes to mind) reversed the usual escalation of attentional blinks with aging, even doing better than another group taken entirely from a younger population.

Perhaps, the German researchers speculate, the nonreactive open awareness—simply noticing and allowing whatever comes into the mind "just to be" rather than following a chain of thoughts about it—becomes a cognitive skill that transfers over to registering a target like the letters and numbers on the blink test without getting caught up in it. That leaves their attention ready for the next target in the sequence—a more efficient way to witness the passing world.

Once the attentional blink had been shown to be reversible, Dutch scientists wondered, What's the minimum training that still lessens the blink? They taught people who had never meditated before how to monitor their mind, using a version of mindfulness.[11] The training sessions lasted just seventeen minutes, after which the volunteers were tested on the attentional blink. They blinked less than a comparison group, who had been taught a focusing meditation that had no effect on this mental skill.

THE MULTITASKING MYTH

We all suffer from the digital-age version of life's "full catastrophe": incoming emails, pressing texts, phone messages, and more, storming in all at once—not to mention the Facebook posts, Instagrams, and all such urgent memos from our personal universe of social media. Given the ubiquity of smartphones and such devices, people today seem to take in far more information than they did before the digital age.

Decades before we began to drown in a sea of distractions, cognitive scientist Herbert Simon made this prescient observation: "What information consumes is attention. A wealth of information means a poverty of attention."

Then, too, there are the ways our social connections suffer. Did you ever have the impulse to tell a child to put down her phone and look in the eyes of the person she is talking to? The need for such advice is becoming increasingly common as digital distractions claim another kind of victim: basic human skills like empathy and social presence.

The symbolic meaning of eye contact, of putting aside what we are doing to connect, lies in the respect, care, even love it indicates. A lack of attention to those around us sends a message of indifference. Such social norms for attention to the people we are with have silently, inexorably shifted.

Yet we are largely impervious to these effects. Many denizens of the digital world, for instance, pride themselves on being able to multitask, carrying on with their essential work even as they graze among all the other incoming channels of what's-up. But compelling research at Stanford University has shown that this very idea is a myth—the brain does not "multitask" but rather switches rapidly from one task (*my work*) to others (*all those funny videos, friends' updates, urgent texts . . .*).[12]

Attention tasks don't really go on in parallel, as "multitasking" implies; instead they demand rapid switching from one thing to the other. And following every such switch, when our attention returns to the original task, its strength has been appreciably diminished. It can take several minutes to ramp up once again to full concentration.

The harm spills over into the rest of life. For one, the inability to filter out the noise (all those distractions) from the signal (what you meant to focus on) creates a confusion about what's important, and so a drop in our ability to retain what matters. Heavy multitaskers, the Stanford group discovered, are more easily distracted in general. And when multitaskers do try to focus on that one thing they have to get done, their brains activate many more areas than just those relevant to the task at hand—a neural indicator of distraction.

Even the ability to multitask efficiently suffers. As the late Clifford Nass, one of the researchers, put it, multitaskers are "suckers for irrelevancy," which hampers not just concentration but also analytic understanding and empathy.[13]

COGNITIVE CONTROL

Cognitive control, on the other hand, lets us focus on a specific goal or task and keep it in mind while resisting distractions, the very abilities multitasking harms. Such steely focus is essential in jobs like air traffic control—where screens can be filled with distractions from the controller's main focus, a given incoming airplane—or just in getting through your daily to-do list.

The good news for multitaskers: cognitive control can be strengthened. Undergrads volunteered to try ten-minute sessions of either focusing on counting their breath or an apt comparison task: browsing *Huffington Post*, Snapchat, or BuzzFeed.[14]

Just three ten-minute sessions of breath counting was enough to appreciably increase their attention skills on a battery of tests. And the biggest gains were among the heavy multitaskers, who did more poorly on those tests initially.

If multitasking results in flabby attention, a concentration workout like counting breaths offers a way to tone up, at least in the short term. But there was no indication that the upward bump in attention would last—the improvement came immediately after the "workout," and so registers on our radar as a state effect, not a lasting trait. The brain's attention circuitry needs more sustained efforts to create a stable trait, as we will see.

Still, even beginners in meditation can sharpen their attention skills, with some surprising benefits. For instance, researchers at the University of California at Santa Barbara gave volunteers an eight-minute instruction of mindfulness of their breath, and found that this short focusing session (compared to reading a newspaper or just relaxing) lessened how much their mind wandered afterward.[15]

While that finding is of interest, the follow-up was even more compelling. The same researchers gave volunteers a two-week course in mindfulness of breathing, as well as of daily activities like eating, for a total of six hours, plus ten-minute booster sessions at home daily.[16] The active control group studied nutrition for the same amount of time. Again, mindfulness improved concentration and lessened mind-wandering.

A surprise: mindfulness also improved working memory—the holding in mind of information so it can transfer into long-term memory. Attention is crucial for working memory; if we aren't paying attention, those digits won't register in the first place.

This training in mindfulness occurred while the students in the study were still in school. The boost to their attention and working memory may help account for the even bigger surprise: mindfulness upped their scores by more than 16 percent on the GRE, the entrance exam for grad school. Students, take note.

Another way cognitive control helps us is in managing our impulses, technically known as "response inhibition." As we saw in chapter five, "A Mind Undisturbed," in Cliff Saron's study the training upped a meditator's ability to inhibit impulse over the course of three months and, impressively, stayed strong in a five-month follow-up.[17] And better impulse inhibition went along with a self-reported uptick in emotional well-being.

META-AWARENESS

When we did our first vipassana courses in India, we found ourselves immersed hour after hour in noting the comings and goings of our own mind, cultivating stability by simply noticing rather than following where those thoughts, impulses, desires, or feelings would have us go. This intensive attention to the movements of our mind boils down to pure meta-awareness.

In meta-awareness it does not matter what we focus our attention on, but rather that we recognize awareness itself. Usually what we perceive is a figure, with awareness in the background. Meta-awareness switches figure and ground in our perception, so awareness itself becomes foremost.

Such awareness of awareness itself lets us monitor our mind without being swept away by the thoughts and feelings we are noticing. "That which is aware of sadness is not sad," observes philosopher Sam Harris. "That which is aware of fear is not fearful. The moment I am lost in thought, however, I'm as confused as anyone else."[18]

Scientists refer to brain activity reflecting our conscious mind and its mental doings as "top-down." "Bottom-up" refers to what goes on in the mind largely outside awareness, in what technically is the "cognitive unconscious." A surprising amount of what we think is from the top down is actually from the bottom up. We seem to impose a top-down gloss on our awareness, where the thin slice of the cognitive unconscious that comes to our attention creates the illusion of being the entirety of mind.[19]

We remain unaware of the much vaster mental machinery of bottom-up processes—at least in the conventional awareness of our everyday life. Meta-awareness lets us see a larger swath of bottom-up operations.

Meta-awareness allows us to track our attention itself—noticing, for example, when our mind has wandered off from something we want to focus on. This ability to monitor the mind without getting swept away introduces a crucial choice point when we find our mind has wandered: we can bring our focus back to the task at hand. This simple mental skill undergirds a huge range of what makes us effective in the world—everything from learning to realizing we've had a creative insight to seeing a project through to its end.

There are two varieties of experience: the "mere awareness" of a thing, which our ordinary consciousness gives us, versus knowing you are aware of that thing—recognizing awareness itself, without judgment or other emotional reactions. For example, we typically watch an engrossing movie so swept away by the plot that we've lost awareness of being in a theater with all its surroundings. But we also can watch a movie attentively while maintaining a background awareness of being in the theater watching a movie. This background awareness doesn't diminish our appreciation and involvement in the movie—it's simply a different mode of awareness.

At the movies the person next to you with a bag of popcorn could be making crunching noises that you tune out but which nevertheless register in your brain. During such unconscious mental processing, activity lessens in a key cortical area, the dorsolateral prefrontal cortex, or DLPFC for short. As you become more aware of being aware, the DLPFC becomes more active.

Consider unconscious bias, the prejudices we hold but do not believe we have at all (as mentioned in chapter six, "Primed for Love"). Meditation can both enhance the function of the DLPFC and lessen unconscious bias.[20]

Cognitive psychologists test meta-awareness by giving people mental tasks so challenging that errors are inevitable, and then tracking the number of such errors—and whether the person notices there might have been an error (that's the meta-awareness angle). These tasks are purposely devilish, designed and calibrated to ensure that whoever takes it will make a certain percentage of mistakes, and, what's more, that their confidence in their responses will vary.

Imagine, for instance, having 160 words whiz by in succession for 1.5 seconds each. Then you see another set of 320 words, half of which you've seen before in that zippy presentation. You have to press one of two buttons to tell if you think the word you see in the second set was in the previous list, or not. Then you rate your confidence in your accuracy for each word, a metric for meta-awareness to the degree you both have confidence in and make the correct response.

Psychologists at UC–Santa Barbara used such a mental challenge with people learning mindfulness for the first time, as well as a group who had a course in nutrition.[21] Meta-awareness improved in the meditation group, but not a whit in those taking the nutrition class.

WILL IT LAST?

Amishi Jha's lab tested the effect of an intensive mindfulness retreat where people meditated for more than eight hours each day for a month.[22] The retreat boosted participants' "alerting," a vigilant state of readiness to respond to whatever you encounter. But although in a previous study she had found an improvement in orienting with beginners in a brief course of mindfulness, surprisingly, these retreat participants showed no such boost.

This nonfinding represents important data if we are to get a full picture of how meditation does and does not matter. It helps us get a portrait of how various aspects of attention change—or do not—with different types of meditation, and at differing levels.

Some changes might occur right away, while others take longer: while orienting may budge initially and then stall, alerting seems to improve with practice. And, we suspect, meditation sustained over time might be needed to maintain such shifts in attention, lest they fade.

About the time when Richie was doing his Harvard research on signal-to-noise shifts in meditators, cognitive scientists such as Anne Treisman and Michael Posner pointed out that "attention" represents too gross a concept. Instead, they argued, we should look at various subtypes of attention, and the neural circuitry each involves. Meditation, the findings now show, seems to enhance many of these subtypes, though we don't yet have the full picture. Amishi's results tell us that picture will be nuanced.

A word of caution: while some aspects of attention improve after just a few hours (or, it seems, minutes) of practice, this by no means indicates those improvements will last. We're skeptical that quickie, one-time interventions matter much after any temporary improvements fade. There is no evidence, for instance, that the erasure of the attentional blink induced by seventeen minutes of mindfulness will make a detectable difference mere hours later, after that state wanes. The same holds for those ten-minute mindfulness sessions that reversed the erosion of focus from multitasking. We suspect that unless you continue the practice every day, multitasking will still weaken your focus.

Our hunch would be that pushing a neural system like attention in a lasting way requires not just these short trainings and continued daily practice, but also intensive booster sessions, as was the case with those who were at the shamatha retreat and then were tested five months later in Cliff Saron's study. Otherwise the brain's wiring will regress to its previous status: a life of distraction punctuated with periods of concentration.

Even so, it's encouraging that such short doses of meditation improve attention. The fact that these improvements come with such rapidity confirms William James's conjecture that sharper attention can be cultivated. Today there are meditation centers in Cambridge no more than a fifteen-minute walk from where William James once lived. Had they been there during his lifetime, and had William James practiced at one, he would no doubt have found his missing education par excellence.

IN A NUTSHELL

Meditation, at its root, retrains attention, and different types boost varying aspects of attention. MBSR strengthens selective attention, while long-term vipassana practice enhances this even more. Even five months after the three-month shamatha retreat, meditators had enhanced vigilance, the ability to sustain their attention. And the attentional blink lessened greatly after three months on a vipassana retreat—but the beginnings of this lessening also showed up after just seventeen minutes of mindfulness in beginners, no doubt a transitory state for the newcomers, and a more lasting trait for those retreatants. That same practice-makes-perfect maxim likely applies to several other quickie meditations: just ten minutes of mindfulness overcame the damage to concentration from multitasking—at least in the short term; only eight minutes of mindfulness lessened mind-wandering for a while. About ten hours of mindfulness over a two-week period strengthened attention and working memory—and led to substantial improved scores on the graduate school entrance exam. While meditation boosts many aspects of attention, these are short-term gains; more lasting benefits no doubt require ongoing practice.

8

Lightness of Being

Back to Richie on his retreat in Dalhousie with S. N. Goenka. A revelation came to Richie on the seventh day, during the Hour of Stillness, which begins with a vow not to make a single voluntary movement, no matter how excruciating your discomfort.

Almost from the start of that endless hour Richie's usual ache in his right knee, now intensified by the no-moving vow, went from pulsating jolts to torture. But then, just as the pain reached the unbearable point, something changed: his awareness.

Suddenly, what had been pain disappeared into a collection of sensations—tingling, burning, pressure—but his knee no longer hurt. The "pain" dissolved into waves of vibrations without a trace of emotional reactivity.

Focusing on just the sensations meant completely reappraising the nature of hurting: instead of fixating on the pain, the very notion of pain deconstructed into raw sensations. What went missing was just as critical: the psychological resistance to, and negative feelings about, those sensations.

The pain had not vanished, but Richie had changed his *relationship* to it. There was just raw sensation—not *my* pain, along with the usual stream of angst-ridden thoughts.

Though while we sit we ordinarily are oblivious to our subtle shifts in posture and the like, these small movements relieve stress that's building in our body. When you don't move a muscle, that stress can build into excruciating pain. And if, like Richie, you are scanning those sensations, a remarkable shift in your relationship to your own experience can occur where the feeling of "pain" melts away into a mélange of physical sensations.

In that hour Richie, with his science background, realized in his most personal reality that what we label as "pain" is a joining together of myriad constituent somatic sensations from which the label arises. With his newly altered perception, "pain" became just an idea, a mental label that puts a conceptual veneer over what arises from a motley coincidence of sensations, perceptions, and resistant thoughts.

This was a vivid taste for Richie of how much mental activity is going on in our mind "under the hood," and about which we are oblivious. He understood that our experience is not based on the direct apperception of what is happening, but to a great extent upon our expectations and projections, the habitual thoughts and reactions that we have learned to make in response, and an impenetrable sea of neural processes. We live in a world our minds build rather than actually perceiving the endless details of what is happening.

This led Richie to a scientific insight: that consciousness operates as an integrator, gluing together a vast amount of elementary mental processes, most of which we are oblivious to. We know their eventual product—*my pain*—but typically have no awareness of the countless elements that combine into that perception.

While that understanding has become a given in cognitive science today, back in the days of the Dalhousie retreat there was no such understanding. Richie had no inkling apart from his own transformation in awareness.

During the first days of the retreat Richie would shift his position now and then to relieve the discomfort in his knees or back. But after that no-moving perceptual breakthrough, Richie could be still as a rock during marathon sessions of up to three hours or longer. With this radical inner shift, Richie felt a sense that he could sit through anything.

Richie saw that if we actually paid attention in the right way to the nature of our experience, it would change dramatically. The Hour of Stillness shows that every waking moment of our lives, we construct our experience around a narrative where we are the star—and that we can deconstruct that story we center on ourselves by applying the right kind of awareness.

HOW OUR BRAIN CONSTRUCTS OUR SELF

Marcus Raichle was surprised—and troubled. Raichle, a neuroscientist at Washington University in St. Louis, had been doing pioneering brain studies to identify which neural areas were active during various mental activities. To do this kind of research back in 2001, Raichle used a strategy common at the time: comparing the active task to a baseline where the participant was doing "nothing." What troubled him: during highly demanding cognitive tasks—like counting backward by 13s from the number 1,475—there were a set of brain regions that *deactivated*.

The standard assumption was that such an effortful mental job would always increase activation in brain areas. But the *deactivation* Raichle found was a systematic pattern, one that accompanies the shift from the resting baseline of doing "nothing" to doing any kind of mental task.

In other words, while we're doing nothing there are brain regions that are highly activated, even more active than those engaged during a difficult cognitive task. While we are working at a mental challenge like tricky subtraction, these brain regions go quiet.

His observation confirmed a mystifying fact that had floated around the world of brain science for a while: that although the brain makes up only 2 percent of the body's mass, it consumes about 20 percent of the body's metabolic energy as measured by its oxygen usage, and that rate of oxygen consumption remains more or less constant no matter what we are doing—including nothing at all. The brain, it seems, stays just as busy when we are relaxed as when we are under some mental strain.

So, where are all those neurons, chatting back and forth while we do nothing in particular? Raichle identified a swath of areas, mainly the mPFC (short for *midline of the prefrontal cortex*) and the PCC (*postcingulate cortex*), a node connecting to the limbic system. He dubbed this circuitry the brain's "default mode network."[1]

While the brain engages in an active task, whether math or meditating, the default areas calm down as those essential for that task gear up, and ramp up again when that mental task finishes. This solved the problem of how the brain could maintain its activity level while "nothing" was going on.

When scientists asked people during these periods of "doing nothing" what was going on in their minds, not surprisingly, it was not nothing! They typically reported that their minds were wandering; most often, this mind-wandering was focused on the self—How am *I* doing in this experiment? *I* wonder what they are learning about *me; I* need to reply to Joe's phone message—all reflecting mental activity focused on "I" and "me."[2]

In short, our mind wanders mostly to something about ourselves— *my thoughts, my emotions, my relationships, who liked my new post on my Facebook page*—all the minutiae of our life story. By framing every event in how it impacts ourselves, the default mode makes each of us the center of the universe as we know it. Those reveries knit together our sense of "self" from the fragmentary memories, hopes, dreams, plans, and so on that center on I, me, and mine. Our default mode continually rescripts a movie where each of us stars, replaying particularly favorite or upsetting scenes over and over.

The default mode turns on while we chill out, not doing anything that requires focus and effort; it blossoms during the mind's downtime. Conversely, as we focus on some challenge, like grappling with what's happened to your Wi-Fi signal, the default mode quiets.

With nothing much else to capture our attention, our mind wanders, very often to what's troubling us—a root cause of everyday angst. For this reason, when Harvard researchers asked thousands of people to report their mental focus and mood at random points through the day, their conclusion was that "a wandering mind is an unhappy mind."

This self-system mulls over our life—especially the problems we face, the difficulties in our relationships, our worries and anxieties. Because the self ruminates on what's bothering us, we feel relieved when we can turn it off. One of the great appeals of high-risk sports like rock climbing seems to be just that—the danger of the sport demands a full focus on where to put your hand or foot next. More mundane worries take backstage in the mind.

The same applies to "flow," the state where people perform at their best. Paying full attention to what's at hand, flow research tells us, rates high on the list of what puts us into—and sustains—a joyous state. The self, in its form as mind-wandering, becomes a distraction, suppressed for the time being.

Managing attention, as we saw in the previous chapter, is an essential ingredient of every variety of meditation. When we become lost in thoughts during meditation, we've fallen into the default mode and its wandering mind.

A basic instruction in almost all forms of meditation urges us to notice when our mind has wandered and then return our focus to the chosen target, say, a mantra or our breathing. This moment has universal familiarity on contemplative paths.

This simple mental move has a neural correlate: activating the connection between the dorsolateral PFC and the default mode—a connection found to be stronger in long-term meditators than in beginners.[3] The stronger this connection, the more likely regulatory circuits in the prefrontal cortex inhibit the default areas, quieting the monkey mind—the incessant self-focused chatter that so often fills our minds when nothing else is pressing.

A Sufi poem hints at this shift, speaking of the shift from "a thousand thoughts" to just one: "There is no god but God."[4]

DECONSTRUCTING THE SELF

As fifth-century Indian sage Vasubhandu observed, "So long as you grasp at the self, you stay bound to the world of suffering."

While most ways to relieve us from the burden of self are temporary, meditation paths aim to make that relief an ongoing fact of life—a lasting trait. Traditional meditative paths contrast our everyday mental states—that stream of thoughts, many laden with angst, or to-do lists that never end—with a state of being free of these weights. And each path, in its particular terms, sees lightening our sense of self as the key to such inner freedom.

When the pain in Richie's knee shifted from excruciating to suddenly bearable, there was a parallel shift in how he identified with it. It was no longer "his" pain; the sense of "mine" had evaporated.

Richie's hour of utter stillness offers a glimpse of how our ordinary "self" can reduce to an optical illusion of the mind. As this keen observation gains strength, at some point our very sense of a solid self breaks down. This shift in how we experience ourselves—our pain and all that we attach to it—points to one of the main goals of all spiritual practice: lightening the system that builds our feelings of I, me, and mine.

The Buddha, in telling of this very insight, likened the self to a chariot, a concept that arises when wheels, platform, yoke, and so on are put together—but which does not exist save as these parts in combination. To update the metaphor, there is no "car" in the tires, nor the dashboard or the steel shell of its body—but put all these together with the multitude of other parts, and what we think of as a car manifests.

In the same way, cognitive science tells us, our sense of self emerges as a property of the many neural subsystems that thread together, among other streams, our memories, our perceptions, our emotions, and our thoughts. Any of those alone would be insufficient for a full sense of our self, but in the right combination we have the cozy feel of our unique being.

Meditative traditions of all kinds share one goal: letting go of the constant grasping—the "stickiness" of our thoughts, emotions, and impulses—that guides us through our days and lives. Technically called "dereification," this key insight has the meditator realize that thoughts, feelings, and impulses are passing, insubstantial mental events. With this insight we don't have to believe our thoughts; instead of following them down some track, we can let them go.

As Dōgen, founder of the Soto school of Zen, instructed, "If a thought arises, take note of it and then dismiss it. When you forget all attachments steadfastly, you will naturally become zazen itself."

Many other traditions see lightening the self as the path to inner freedom. We've often heard the Dalai Lama talk about "emptiness," by which he means the sense in which our "self"—and all seeming objects in our world—actually emerge from the combination of their components.

Some Christian theologians use the term *kenosis* for the emptying of self, where our own wants and needs diminish while our openness to the needs of others grows into compassion. As a Sufi teacher put it, "When occupied with self, you are separated from God. The way to God is but one step; the step out of yourself."[5]

Such a step out of the self, technically speaking, suggests weakening activation of the default circuitry that binds together the mosaic of memories, thoughts, impulses, and other semi-independent mental processes into the cohesive sense of "me" and "mine."

The stuff of our lives becomes less "sticky" as we shift into a less attached relationship toward all that. At the higher reaches of practice, mind training lessens the activity of our "self." "Me" and "mine" lose their self-hypnotic power; our concerns become less burdensome. Though the bill still must be paid, the lighter our "selfing," the less we anguish about that bill and the freer we feel. We still find a way to pay it, but without the extra load of emotional baggage.

While almost every contemplative path holds lightness of being as a primary aim, paradoxically, very little scientific research speaks to this goal. Our reading of the meager studies done so far suggests there may be three stages in how meditation leads to greater selflessness. Each of these stages uses a different neural strategy to quiet the brain's default mode, and so free us a bit from the grip of the self.

THE DATA

David Creswell, now at Carnegie Mellon University, was another young scientist whose interest in meditation was nurtured by attending the Mind and Life Summer Research Institute. To assess the early stage, found among meditation novices, Creswell's group measured brain activity in people who volunteered for a three-day intensive course in mindfulness.[6] The volunteers had never meditated before, but in this mindfulness course they learned that if you are lost in some personal melodrama (a favorite theme of the default mode), you can voluntarily drop it—you can name it, or shift your attention to watching your breath or to bare awareness of the present moment. All of these are active interventions, efforts to quiet the monkey mind.

Such efforts heighten activity in the dorsolateral prefrontal area, a key circuit for managing the default mode. As we've seen, this area springs into action anytime we intentionally attempt to quiet our agitated mind—for instance, when we try to think of something more pleasant than some upsetting encounter that keeps running over and over in our mind.

Three days of practicing these mindfulness methods led to increased connections between this control circuitry and the default zone's PCC, a primary region for self-focused thought. Novices in meditation, this suggests, keep their mind from wandering by activating neural wiring that can quiet the default area.

But with more experienced meditators, the next phase of downscaling the self adds lessened activity in key sections of the default mode—a loosening of the mechanics of self—while the heightened connections with control areas continue. A case in point: researchers led by Judson Brewer, then at Yale University, (and who has been on the faculty at the SRI) explored brain correlates of mindfulness practice, comparing highly experienced meditators (averaging around 10,500 lifetime hours) with novices.[7]

During the meditation practice, all those tested were encouraged to distinguish between simply noting the identity of an experience (*itching is occurring*, say) and identifying with it (*I* itch)—and then to let go. This distinction seems a crucial step in loosening the self, by activating meta-awareness—a "minimal self" that can simply notice the itch rather than bring it into our story line, *my itch*.

As mentioned, when we are watching a movie and are lost in its story, but then notice that we are in a movie theater watching a film, we have stepped out of the movie's world into a large frame that includes the movie but goes beyond. Having such meta-awareness allows us to monitor our thoughts, feelings, and actions; to manage them as we like; and to inquire into their dynamics.

Our sense of self gets woven in an ongoing personal narrative that threads together disparate parts of our life into a coherent story line. This narrator resides mainly in the default mode but brings together inputs from a broad range of brain areas that in themselves have nothing to do with the sense of self.

The seasoned meditators in the Brewer study had the same strong connection between the control circuit and the default mode seen in beginners, but in addition had less activation within the default mode areas themselves. This was particularly true when they practiced loving-kindness meditation—a corroboration of the maxim that the more we think of the well-being of others, the less we focus on ourselves.[8]

Intriguingly, the long-term meditators seemed to have roughly the same lessened connectivity in the default mode circuitry while they just rested before the test as they displayed during mindfulness. That's a likely trait effect and a good sign: these meditators intentionally train to be as mindful in their daily lives as during meditation sessions. The same lessened connectivity compared to nonmeditators was found by brain researchers in Israel studying long-term mindfulness meditators, who had on average around 9,000 hours of practice under their belts.[9]

Further indirect evidence for this change in long-term meditators comes from a study at Emory University of seasoned Zen meditators (three years–plus practice, but lifetime hours unknown) who, compared to controls, seemed to show less activity in parts of the default area while focusing on their breath during brain scans. The bigger this effect, the better they did on a test of sustained attention outside the scanner, suggesting a lasting drop in mind-wandering.[10] Finally, a small but suggestive study of Zen meditators at the University of Montreal found lessened default area connectivity while just resting among Zen meditators (with an average 1,700 hours of practice) compared to a group of volunteers trained in zazen for just one week.[11]

There's a theory that what captures our attention signifies an attachment, and the more attached we are, the more often we'll be so captivated. In an experiment testing this premise, a group of volunteers and one of seasoned meditators (4,200 hours) were told they would get money whenever they recognized certain geometric shapes within an array.[12] That was, in a sense, the creation of a mini-attachment. Then, in a later phase, when they were told to simply focus on their breath and ignore those shapes, the meditators were less distracted by them than were the control group.

Along these lines, Richie's group found that meditators who had an average 7,500 lifetime hours, compared to people their own age, had decreased gray matter volume in a key region: the nucleus accumbens.[13] This was the only brain region showing a difference in brain structure compared to age-matched controls. A smaller nucleus accumbens diminishes connectivity between these self-related regions and the other neural modules that ordinarily orchestrate to create our sense of self.

This is a bit of a surprise: the nucleus accumbens plays a large role in the brain's "reward" circuits, a source of pleasurable feelings in life. But this is also a key area for "stickiness," our emotional attachments, and addictions—in short, what ensnares us. This decrease in gray-matter volume in the nucleus accumbens may reflect a diminished attachment in the meditators, particularly to the narrative self.

So, does this change leave meditators cold and indifferent? The Dalai Lama and other highly seasoned practitioners come to mind—like those who came to Richie's lab, most of whom tend toward joyousness and warmth.

Meditation texts describe long-term practitioners achieving an ongoing compassion and bliss, but with "emptiness," in the sense of no attachment. For instance, Hindu contemplative paths describe *vairagya,* a later stage of practice where attachments drop away—renunciation, in this sense, happens spontaneously rather than through force of will. And with this shift emerges an alternate source of delight in sheer being.[14]

Could this indicate a neural circuit that brings a quiet enjoyment, even as our nucleus accumbens–based attachments wane? We will see just such a possibility in chapter twelve, "Hidden Treasure," from brain studies of advanced yogis.

Arthur Zajonc, the second president of the Mind and Life Institute, and a quantum physicist and philosopher to boot, once said that if we let go of grasping, "we become more open to our own experience, and to other people. That openness—a form of love—lets us more easily approach other people's suffering."

"Great souls," he added, "seem to embody the ability to engage suffering and handle it without collapse. Letting go of grasping is liberating, creating a moral axis for action and compassion."[15]

A THIEF IN AN EMPTY HOUSE

Ancient meditation manuals say letting go of these thoughts is, at first, like a snake uncoiling itself; it takes some effort. Later, though, whatever thoughts come to mind are like a thief entering an empty house: there's nothing to do, so they just leave.

This segue from at first making an effort to later effortlessness seems a universal, though little-known, theme in meditation paths. Common sense tells us that learning any new skill takes hard work at first and becomes progressively easier with practice. Cognitive neuroscience tells us this shift to effortlessness marks a neural transition in habit mastery: the prefrontal areas no longer make an effort to do the work, as the basal ganglia lower in the brain can take over—a neural mode that marks effortlessness.

Effortful practice at the early stages of meditation activates prefrontal regulatory circuits. However, the later shift to effortless practice might go along with a different dynamic: lessened connectivity among the various nodes of the default circuitry, and lessened activity in the PCC as effortful control is no longer needed—the mind at this stage is truly beginning to settle and the self-narrative is much less sticky.

That was found in another study by Judson Brewer, where seasoned meditators reported their experience in the moment, allowing scientists to see what brain activity correlated with it. When the meditators showed decreased activity in their PCC, they reported feelings like "undistracted awareness" and "effortless doing."[16]

In the scientific study of any skill that people practice, from dentistry to chess, when it comes to sorting out the duffers from the pros, lifetime hours of practice are gold. A pattern of high effort at the start segueing into less effort along with more proficiency in a task shows up in experts as diverse as swimmers and violinists. And as we've seen here, the brains of those with the most hours of meditation showed little effort in keeping their focus one-pointed, even despite compelling distractions, while those with fewer lifetime hours put in more effort. And at the very start, beginners showed an increase in biological markers of mental effort.[17]

The rule of thumb: the brain of a novice works hard while that of the expert expends little energy. As we master any activity, the brain conserves its fuel by putting that action on "automatic"; cuing up that activity shifts from top-of-the-brain circuits to the basal ganglia far below the neocortex. We've all accomplished the hard-at-first to no-sweat transition when we learned to walk—and as we've mastered every other habit since. What at first demands attention and exertion becomes automatic and effortless.

At the third and final stage of letting go of self-referencing, we conjecture, the control circuitry's role drops away, as the main action shifts to looser connectivity in the default mode, the home of the self. Brewer's group found such a decrease.

With a spontaneous shift to effortlessness comes a change in the relationship to the self: it's not so "sticky" anymore. The same sorts of thoughts can arise in your mind, but they are lighter: not so compelling, with less emotional oomph, and so float away more easily. This, at any rate, reflects what we hear from the advanced yogis studied in the Davidson lab, as well as from classic meditation manuals.

But we have no data on this point, which remains a ripe research question. And what that future research might find could be surprising—for example, with this shift in relationship to the self, we may see change not so much in the currently known neural "self-systems" but rather in other circuitry yet to be discovered.

Lessening the grip of the self, always a major goal of meditation practitioners, has been oddly ignored by meditation researchers, who perhaps understandably focus instead on more popular benefits like relaxation and better health. And so, a key goal of meditation—

selflessness—has only thin data, while other benefits, like health improvements, are heavily researched, as we will see in the next chapter.

A LACK OF STICKINESS

Richie once saw tears begin to stream down the Dalai Lama's face as he heard about a tragic situation in Tibet—the latest self-immolation among Tibetans protesting the Communist Chinese occupation of their land.

And then, a few moments later, the Dalai Lama noticed someone in the room doing something funny and he began laughing. There was no disrespect for the tragedy that brought him to tears, but rather, a buoyant and seamless transition from one emotional note to the other.

Paul Ekman, a world expert on emotions and their expression, says this remarkable affective flexibility in the Dalai Lama struck him as exceptional from their very first meeting. The Dalai Lama reflects in his own demeanor the emotions he feels from one person, and then immediately drops that feeling as the next moment brings him another emotional reality.[18]

The Dalai Lama's emotional life seems to include a remarkably dynamic range of strong and colorful emotions, from intense sadness to powerful joy. His rapid, seamless transitions from one to another are particularly unique—this swift shifting betokens a lack of stickiness.

Stickiness seems to reflect the dynamics of the emotional circuitry of the brain, including the amygdala and the nucleus accumbens. These regions very likely underlie what traditional texts see as the root causes of suffering—attachment and aversion—where the mind becomes fixated on wanting something that seems rewarding or on getting rid of something unpleasant.

The stickiness spectrum runs from being utterly stuck, unable to free ourselves from distressing emotions or addictive wants, to the Dalai Lama's instant freedom from any given affect. One trait that emerges from living without getting stuck seems to be an ongoing positivity, even joy.

When the Dalai Lama once was asked what had been the happiest point in his life, he answered, "I think right now."

IN A NUTSHELL

The brain's default mode activates when we are doing nothing that demands mental effort, just letting our mind wander; we hash over thoughts and feelings (often unpleasant) that focus on ourselves, constructing the narrative we experience as our "self." The default mode circuits quiet during mindfulness and loving-kindness meditation. In early stages of meditation this quieting of the self-system entails brain circuits that inhibit the default zones; in later practice the connections and activity within those areas wane.

This quieting of the self-circuitry begins as a state effect, seen during or immediately after meditation, but with long-term practitioners it becomes an enduring trait, along with lessened activity in the default mode itself. The resulting decrease in stickiness means that self-focused thoughts and feelings that arise in the mind have much less "grab" and decreasing ability to hijack attention.

9

Mind, Body, and Genome

When Jon Kabat-Zinn first developed MBSR at the University of Massachusetts Medical Center in Worcester, he started slowly, talking one by one to physicians there. He invited them to refer their patients who had to endure chronic conditions like untreatable pain—those considered medical "failures," because even narcotics didn't help—or who had to manage lifelong conditions like diabetes or heart disease. Jon never claimed he could cure such diseases. His mission: improve the quality of patients' lives.

Surprisingly, perhaps, Jon met with almost no resistance from physicians. Right from the start, key clinic directors (primary care, pain, orthopedics) were willing to send such patients to what Jon at the time called the Stress Reduction and Relaxation Program, based in a basement room borrowed from the physical therapy department.

Jon led sessions there just a few days a week. But as word spread of patients praising the method for making their lives with an incurable condition more bearable, the program flourished and, in 1995, expanded into the Center for Mindfulness in Medicine, Health Care, and Society to house its research, clinical, and professional educational programs. Today hospitals and clinics around the world offer MBSR, one of the fastest-growing kinds of meditation practice, and by now the approach with the strongest empirical evidence of its benefits. Beyond health care, MBSR has become ubiquitous, spearheading the popular mindfulness movement in psychotherapy, education, and even business.

Now taught at most academic medical centers in North America and in many parts of Europe, MBSR offers a standard program that makes it appealing for scientific study. To date there are more than six hundred published studies of the method, revealing a wide variety of benefits—and some instructive caveats.

For instance, medicine sometimes falters when it comes to treating chronic pain. Aspirin and other over-the-counter pain medications can have too many side effects to be used daily for years; steroids offer temporary relief but again with sometimes harmful side effects; and opioids have proven too addictive to be used widely. MBSR, however, can help without such drawbacks, since there are usually no negative side effects of mindfulness practice, and if practiced following the eight-week MBSR program, can continue to help people live well with chronic conditions and with stress-related disorders that will not necessarily get better on their own or with conventional medical treatment. A key element for long-term benefit is the continuity of practice, and despite MBSR's long history, we still have virtually no good information on the extent to which those who have taken an MBSR course continue to engage in formal practice in the years following their initial training.

Take debilitating pain in the elderly. One of the most feared impacts of growing old is losing independence due to troubles with mobility from pain in arthritic hips, knees, or spine. In well-designed research with elderly pain sufferers, MBSR proved highly effective both in reducing how much pain people felt and how disabled they became as a result.[1] Their lowered pain levels lasted into a six-month follow-up.

As in all MBSR programs, participants were urged to continue a daily practice at home. Having a method they can use on their own to ease their pain gave these patients a sense of "self-efficacy," a feeling that they can control their destiny to some extent. This in itself helps patients live better with pain that won't go away.

When Dutch researchers analyzed dozens of studies on mindfulness as a pain treatment, they concluded this approach was a good alternative to purely medical treatment.[2] Even so, no research so far has found that meditation produces clinical improvements in chronic pain by removing the biological cause of the pain—the relief comes in how people relate to their pain.

Fibromyalgia offers an instructive case in point. This malady presents a medical mystery: there are no known biological explanations for the chronic pain, fatigue, stiffness, and insomnia that typify this debilitating disorder. The one exception seems to be impairment in regulating heart function (though this, too, is debated). One gold standard study that used MBSR with women who suffered from fibromyalgia failed to find any impact on cardiac activity.[3]

Even so, another well-designed study found that MBSR brought significant improvements in psychological symptoms, such as how much stress fibromyalgia patients felt, and lessened many of their subjective symptoms.[4] The more often they used MBSR on their own, the better they did. Still, there was no change in the patients' physical functioning or in a key stress hormone, cortisol, which stayed at high levels. The patients' relationship to their pain changed for the better with MBSR—but not the underlying biology causing the pain itself.

Should someone with a disorder like chronic pain or fibromyalgia try MBSR, or meditation of any kind? Depends who you ask.

Medical researchers, in endless pursuit of definitive outcomes, have one set of criteria; patients have quite another. While doctors may want to see hard data showing medical improvements, patients just want to feel better, especially if there's little to be done to relieve their clinical condition. From a patient's viewpoint, then, mindfulness offers a path to relief—even as medical research tells doctors the evidence is not clear when it comes to reversing the biological cause of the pain.

Though patients may find relief from pain after they have gone through the eight-week MBSR course, many drop the practice after a while. That may be why several studies have found good results for patients immediately after they take MBSR, but less so in six-month follow-ups. So—as Jon will tell you—the key to a lifetime relatively free from the experience of pain, both physical and emotional, is continuing one's mindfulness practice day after day in the following months, years, and decades.

WHAT THE SKIN REVEALS

Our skin offers a surprising window on how stress impacts our health. As a barrier tissue in direct contact with foreign agents from the world around us (as are the gastrointestinal tract and our lungs), the skin is part of the body's first line of defense against invading germs. Inflammation signals a biological defensive maneuver that walls off infection from healthy tissue so it won't spread. A red, inflamed patch signals that the skin has attacked a pathogen.

The degree of inflammation in the brain and body play a big role in how severe a disease like Alzheimer's, asthma, or diabetes will be. Stress, though often psychological, worsens inflammation, apparently part of an ancient biological response to warnings of danger that marshals the body's resources for recovery. (Another signal of that response: how you just want to rest when you get the flu.) While the threats that trigger this response in prehistory were physical, like something that could eat us, these days the triggers are psychological—an angry spouse, a snarky tweet. Yet the body's reactions are the same, including emotional upset.

Human skin has an unusually large number of nerve endings (about five hundred per square inch), each a pathway for the brain to send signals for what's called "neurogenic," or brain-caused, inflammation. Skin specialists have long observed that life's stress can cause neurogenic flare-ups of inflammatory disorders like psoriasis and eczema. This makes the skin an appealing lab for studying how upsets impact our health.

Turns out the nerve pathways that let the brain signal the skin to inflame are sensitive to capsaicin, the chemical that makes chilies "hot." Richie's lab used this novel fact to create carefully controlled patches of inflammation, to see how stress would increase, or meditation muffle, this reaction. Meanwhile, Melissa Rosenkranz, a scientist in the lab, invented a clever way to assay the chemicals that induce inflammation, by creating artificial (and painless) blisters in the inflamed area that would fill with fluid.

The blisters were created in a contraption Melissa built that uses a vacuum system to raise the first layer of skin in small circular areas over the course of forty-five minutes. When done slowly the method is quite painless, hardly noticed by the participants. Tapping that fluid allowed measuring levels of pro-inflammatory cytokines, the type of proteins that directly cause those red patches.

Richie's lab compared a group who were taught MBSR with another who went through HEP (the active control treatment) as they endured the Trier ordeal—a dispiriting job interview, followed by a tough math workout—a sure way to trigger the pandemonium of the stress response.[5] More specifically, the brain's threat radar, the amygdala, signals the HPA axis (that's the hypothalamic-pituitary-adrenal circuitry, if you must know) to release epinephrine, an important freeze-fight-or-flight brain chemical, along with the stress hormone cortisol, which in turn raises the body's energy expenditure to respond to the stressor.

In addition, in order for the body to ward off bacteria in wounds, pro-inflammatory cytokines increase blood flow to the area to supply immune products that gobble up foreign substances. The resulting inflammation in turn signals the brain in ways that activate several neural circuits, including the insula and its extensive connections throughout the brain. One of the areas triggered by messages from the insula is the anterior cingulate cortex (ACC), which modulates inflammation and also connects our thoughts and feelings and controls autonomic activity, including heart rate. Richie's group discovered that when the ACC activates in response to an allergen, people with asthma will have more attacks twenty-four hours later.[6]

Back to the inflammation study. There were no differences in the two groups' subjective reports of distress, nor in their levels of the cytokines that trigger inflammation, nor in cortisol, that hormonal precursor of diseases made worse by chronic stress, like diabetes, hardening of the arteries, and asthma.

But the MBSR group did better on an unfudgeable test: participants had a significantly smaller patch of inflammation after the stress test, and their skin was more resilient, healing faster. That difference held even four months later.

Although the subjective benefits of MBSR, and some of the biological ones, do not seem unique, this impact on inflammation certainly seems to be. Those who engaged in their MBSR practices for thirty-five minutes or more at home daily, compared to those doing HEP, showed a greater decrease in pro-inflammatory cytokines, the proteins that trigger the red patch. This, intriguingly, supports an early finding by Jon Kabat-Zinn and some skin specialists that MBSR can help speed healing from psoriasis, a condition worsened by inflammatory cytokines (but some thirty years on, this remains a study not yet replicated by dermatology researchers).[7]

To get a better idea of how meditation practice might heal such inflammatory conditions, Richie's lab repeated the stress study using highly experienced (around 9,000 lifetime hours of practice) vipassana meditators.[8] Result: the meditators not only found the dreaded Trier test less stressful than did a matched cohort of novices (as we saw in chapter five), but they also had smaller patches of inflammation afterward. Most significant, their levels of the stress hormone cortisol were 13 percent lower than in the controls, a substantial difference that is likely clinically meaningful. And the meditators reported being in better mental health than volunteers matched for age and gender who did not meditate.

Important: these seasoned practitioners were not meditating when these measures were taken—this was a trait effect. Mindfulness practice, it seems, lessens inflammation day to day, not just during meditation itself. The benefits seem to show up even with just four weeks of mindfulness practice (around thirty hours total), as well as with loving-kindness meditation.[9] While those new to MBSR had a mild trend toward lower cortisol, a large drop in cortisol under stress seems to kick in at some point with continued practice. Looks like there's biological confirmation of what meditators say: it gets easier to handle life's upsets.

Constant stress and worry take a toll on our cells, aging them. So do continual distractions and a wandering mind, due to the toxic effects of rumination, where our mind gravitates to troubles in our relationships but never resolves them.

David Creswell (whose research we visited in chapter seven) recruited unemployed job seekers—a highly stressed group—and offered them either a three-day intensive program of mindfulness training or a comparable relaxation program.[10] Blood samples before and after revealed that the meditators, but not those taking relaxation, had reductions in a key pro-inflammatory cytokine.

And, fMRI scans showed, the greater their increase in connectivity between the prefrontal region and the default areas that generate our inner stream of chat, the greater the reductions in the cytokine. Presumably, putting the brakes on destructive self-talk that floods us with thoughts of hopelessness and depression—understandable in the unemployed—also lowered cytokine levels. How we relate to our gloomy self-talk has a direct impact on our health.

HYPERTENSION? RELAX.

The moment you woke up today, were you breathing in or breathing out?

That hard-to-answer question was put to a retreatant by the late Burmese monk and meditation master Sayadaw U Pandita. It bespeaks the extremely conscientious and precise version of mindfulness he was renowned for teaching.

The sayadaw was the direct lineage holder of the great Burmese teacher Mahasi Sayadaw, as well as spiritual guide to Aung San Suu Kyi during her years-long house arrest before she became Burma's head of government. On his occasional trips to the West, Mahasi Sayadaw had instructed many of the best-known teachers in the vipassana world.

Dan had traveled off-season to a rented kids' summer camp in the high desert of Arizona to spend a few weeks under U Pandita's guidance. As Dan later wrote in the *New York Times Magazine*, "The consuming task of my day was to build a precise attention to my breath, noticing every nuance of each inhalation and exhalation: its speed, lightness, coarseness, warmth."[11] The point for Dan: clear the mind, and so, calm the body.

While this retreat was one of a series Dan tried to fit into his yearly calendar in the decades after returning from his graduate school sojourns in Asia, it wasn't just meditation progress he hoped for. Over the fifteen years or so since his last long stay in India, his blood pressure had gotten too high, and Dan hoped this retreat would lower it, at least for a while. His physician had been troubled by readings over 140/90, the lower border of hypertension. And when Dan returned home from retreat, he was pleased to find a reading far below that borderline.

The notion that people could lower blood pressure through meditation largely originated with Dr. Herbert Benson, a Harvard Medical School cardiologist. When we were at Harvard, Dr. Benson had just published one of the first studies on the topic showing meditation seemed to help lower blood pressure.

Herb, as we know him, served on Dan's dissertation committee, and was one of the few faculty members anywhere at Harvard sympathetic to meditation studies. As later research on meditation and blood pressure have shown, he was on the right track.

Take, for example, a well-designed study of African American men, who are at particularly high risk for hypertension, cardiac and kidney disease. Just fourteen minutes of mindfulness practice in a group who already suffered from kidney disease lowered the metabolic patterns that, if sustained year after year, lead to these diseases.[12]

The next step, of course, would be to try mindfulness (or some other variety of meditation) with a similar group, but who had not yet developed a full-blown disease, compare them with a matched group who did something like HEP, and follow them for several years to see if meditation headed off the disease (as we would hope—but let's try this study to see for sure).

On the other hand, when we look at a larger set of studies the news here is mixed. In a meta-analysis of eleven clinical studies where patients with conditions like heart failure and ischemic heart disease were randomly assigned to meditation training or a comparison group, results were, in the words of the researchers, "encouraging" but not conclusive.[13] As usual, the meta-analysis called for larger and more rigorous studies.

There's a growing body of research here but a meager yield when we look for well-designed studies. Most have randomized wait-list controls, which is good, but usually lack an active control group, which would be best. Only with an active control do we know that the benefits are due to the meditation itself rather than to the "nonspecific" impact of having an encouraging instructor and a supportive group.

GENOMICS

"It's just naive," a grant reviewer bluntly told Richie, to think that one will see changes in how genes are expressed during just one day of meditation. Richie had just received the same negative opinion via a review from the National Institutes of Health rejecting his proposal for that exact study.

Some background. After genetic scientists mapped the entire human genome, they realized it wasn't enough to just know if we had a given gene or not. The real questions: Is that gene expressed? Is it manufacturing the protein for which it is designed? And how much? Where is the "volume control" on the gene set?

This meant there was another important step: finding what turns our genes on or off. If we've inherited a gene that gives us a susceptibility to a disease like diabetes, we may never develop the malady if, for example, we have a lifelong habit of getting regular exercise and not eating sugar.

Sugar turns on the genes for diabetes; exercise turns them off. Sugar and exercise are "epigenetic" influencers, among the many, many factors that control whether or not a gene expresses itself. Epigenetics has become a frontier of genomic studies. And Richie thought meditation just might have epigenetic impacts, "down-regulating" the genes responsible for the inflammatory response. As we've seen, meditation seems to do this—but the genetic mechanism for the effect was a complete mystery.

Undeterred by the skeptics, his lab went ahead, assaying changes in the expression of key genes before and after a day of meditation in a group of long-term vipassana practitioners (average of about 6,000 lifetime hours).[14] They followed a fixed eight-hour schedule of practice sessions throughout the day, and listened to tapes of some inspiring talks and guided practices by Joseph Goldstein.

After the day of practice the meditators had a marked "down-regulation" of inflammatory genes—something that had never been seen before in response to a purely mental practice. Such a drop, if sustained over a lifetime, might help combat diseases with onsets marked by chronic low-grade inflammation. As we've said, these include many of the world's major health problems, ranging from cardiovascular disorders, arthritis, and diabetes to cancer.

And this epigenetic impact, remember, was a "naive" idea that countered the then prevailing wisdom in genetic science. Despite assumptions to the contrary, Richie's group had shown that a mental exercise, meditation, could be a driver of benefits at the level of genes. Genetic science would have to change its assumptions about how the mind can help manage the body.

A handful of other studies find that meditation seems to have salutary epigenetic effects. Loneliness, for instance, spurs higher levels of pro-inflammatory genes; MBSR can not only lower those levels— but also lessen the feeling of being lonely.[15] Though these were pilot studies, an epigenetic boost was found in research with two other meditation methods. One is Herb Benson's "relaxation response," which has a person silently repeat a chosen word like *peace* as if it were a mantra.[16] The other is "yogic meditation," where the meditator recites a Sanskrit mantra, at first aloud and then in a whisper, and finally silently, ending with a short deep-breathing relaxation technique.[17]

There are other promising hints for meditation as a force in upgrading our epigenetics. Telomeres are the caps at the end of DNA strands that reflect how long a cell will live. The longer the telomere, the longer the life span of that cell will be.

Telomerase is the enzyme that slows the age-related shortening of telomeres; the more telomerase, the better for health and longevity. A meta-analysis of four randomized controlled studies involving a total of 190 meditators found practicing mindfulness was associated with increased telomerase activity.[18]

Cliff Saron's project found the same effect after three months of intensive practice of mindfulness and compassion meditation.[19] The more present to their immediate experience, and the less mind-wandering during concentration sessions, the greater the telomerase benefit. And a promising pilot study found longer telomeres in women who had an average of four years of regular practice of loving-kindness meditation.[20]

Then there's *panchakarma*, Sanskrit for "five treatments," which mixes herbal medicines, massage, dietary changes, and yoga with meditation. This approach has its roots in Ayurvedic medicine, an ancient Indian healing system, and has become an offering at some upscale health resorts in the United States (and at many lower-cost health spas in India, if you're interested).

A group who went through a six-day panchakarma treatment, compared to another group who were just vacationing at the same resort, showed intriguing improvements in a range of sophisticated metabolic measures that reflect both epigenetic changes and actual protein expression.[21] This means genes are being directed in beneficial ways.

But here's our problem: while there might be some positive health impacts from panchakarma, the mix of treatments makes it impossible to tell how much any one of them, like meditation, was an active agent. The study used five different kinds of interventions together. Such a mishmash (technically, a confound) makes it impossible to tell if the meditation was the active force, or perhaps some herb in the medicine, or a vegetarian diet, or if something else in that mix accounts for the improvements. Benefits accrue—we just don't know why.

There's also the gap between showing improvements at the genetic level and proving meditation has biological effects that matter medically. None of these studies makes that further connection.

In addition, there's the issue of what kind of meditation has which physiological impacts. Tania Singer's group compared concentrating on the breath with loving-kindness and also with mindfulness, looking at how each influenced heart rate and how much effort meditators reported the methods took.[22] The breath meditation was the most relaxing, with loving-kindness and mindfulness both boosting heart rate a bit, a sign these take more effort. Richie's lab had a similar increase in heart rate with highly experienced meditators (more than 30,000 lifetime hours) doing compassion meditation.[23]

While a quicker heartbeat seems a side effect of these warm-hearted meditations—a state effect—when it comes to the breath, the trait payoff goes in the other direction. Science has long known that people with problems like anxiety disorders and chronic pain breathe more quickly and less regularly than most folks. And if you're already breathing fast, you are more likely to trigger a freeze-fight-or-flight reaction when confronting something stressful.

But consider what Richie's lab found when they looked at long-term meditators (9,000 average lifetime hours of practice).[24] Comparing each to a nonmeditator of the same age and sex, the meditators were breathing an average 1.6 breaths more slowly. And this was while they were just sitting still, waiting for a cognitive test to start.

Over the course of a single day that difference in breath rate translates to more than 2,000 extra breaths for the nonmeditators—and more than 800,000 extra breaths over the course of a year. These extra breaths are physiologically taxing, and can exact a health toll as time goes on.

As practice continues and breathing becomes progressively slower, the body adjusts its physiological set point for its respiratory rate accordingly. That's a good thing. While chronic rapid breathing signifies ongoing anxiety, a slower breath rate indicates reduced autonomic activity, better mood, and salutary health.

THE MEDITATOR'S BRAIN

You may have heard the good news that meditation thickens key parts of the brain. The first scientific report of this neural benefit came in 2005 from Sara Lazar, an early grad of Mind and Life's Summer Research Institute, who became a researcher at Harvard Medical School.[25]

Compared with nonmeditators, her group reported, meditators had greater cortical thickness in areas important for sensing inside one's own body and for attention, specifically the anterior insula and zones of the prefrontal cortex.

Sara's report has been followed by a stream of others, many (but not all) reporting increased size in key parts of meditators' brains. Less than a decade later (a very short time given how long such research takes to ramp up, execute, analyze, and report), there were enough brain imaging studies of meditators to justify a meta-analysis, where twenty-one studies were combined to see what held up, what did not.[26]

The results: certain areas of the brain seemed to enlarge in meditators, among them:

- The insula, which attunes us to our internal state and powers emotional self-awareness, by enhancing attention to such internal signals.
- Somatomotor areas, the main cortical hubs for sensing touch and pain, perhaps another benefit of increased bodily awareness.
- Parts of the prefrontal cortex that operate in paying attention and in meta-awareness, abilities cultivated in almost all forms of meditation.
- Regions of the cingulate cortex instrumental in self-regulation, another skill practiced in meditation.
- The orbitofrontal cortex, also part of the circuitry for self-regulation.

And the big news about meditation for older folks comes from a study at UCLA that finds meditation slows the usual shrinkage of our brain as we age: at age fifty, longtime meditators' brains are "younger" by 7.5 years compared to brains of nonmeditators of the same age.[27] Bonus: for every year beyond fifty, the brains of practitioners were younger than their peers' by one month and twenty-two days.

Meditation, the researchers conclude, helps preserve the brain by slowing atrophy. While we doubt that brain atrophy actually can be reversed, we have reason to agree it can be slowed.

But here's the trouble with the evidence so far. That finding on meditation and aging brains was a reanalysis of an earlier study done at UCLA that recruited fifty meditators and fifty people matched for age and sex who had never meditated. The research team made careful images of their brains and found meditators showed greater cortical gyrification (the folding at the top of the neocortex) and so had more brain growth.[28] The longer the meditator had practiced, the more folding.

But as the researchers themselves noted, the findings raise many questions. The specific varieties of meditation practiced among those fifty ranged from vipassana and Zen to kriya and kundalini forms of yoga. These practices can vary greatly in the particular mental skill being deployed by a meditator, for example open presence where anything can come into the mind versus a tight focus on one thing only, or methods that manage breathing versus those that let breathing be natural. Thousands of hours of practice of each of these could well have quite unique impacts, including in neuroplasticity. We don't know from this study what method results in which change—does every kind of meditation lead to the increases that cause more folding or do just a few account for the bulk of it?

This conflation of different kinds of meditation, as though they were all the same (and so have similar brain impacts), pertains also to that meta-analysis. Since the studies included were also a mix of meditation types there's the dilemma that all but a few of the brain-imaging findings are "cross-sectional"—a one-time image of the brain.

The differences could be due to factors like education or exercise, each of which has its own buffering effect on brains. Then there's self-selection: perhaps people with the brain changes reported in these studies choose to stick with meditation, while others do not—maybe having a bigger insula in the first place makes you like meditation more. Each of these alternate potential causes has nothing to do with meditation.

To be fair, the researchers themselves name such drawbacks to their study. But we highlight them here to underline the ways in which a complicated, poorly understood, and tentative scientific finding can radiate out to the general public as an oversimplified message of "meditation builds the brain." The devil, as the saying has it, is in the details.

So now let's consider some promising results from three studies that looked at how just a little meditation practice seemed to have increased volume in parts of the brain, based on differences found before and after trying the practice.[29] Similar results of increases in thickness and the like of appropriate brain areas come from other kinds of mental training like memorization—and neuroplasticity means this is quite possible with meditation.

But here's the big problem with all these studies: they have a very small number of subjects, not enough to reach definitive conclusions. We need many more participants in these studies because of another problem: the brain measures used are relatively squishy, based on statistical analyses of about 300,000 voxels (a voxel is a volume unit, essentially a three-dimensional pixel, each a 1 cubic millimeter hunk of neural geography).

Odds are, a small portion of these 300,000 analyses will show up as statistically "significant," when they are actually random, a problem that diminishes as the number of brains being imaged increases. For now, there's no way to know in these studies if the findings of brain growth are actual or an artifact of the methods used. Another problem: researchers tend to publish their positive findings but not report nonfindings—times they did *not* find any effect.[30]

Finally, brain measures have become more precise and sophisticated since many of these studies were done. We don't know if measurements using the newer, more stringent criteria would yield the same findings. Our hunch is that better studies will reveal positive changes in brain structure with meditation, but it's too early to say. We're waiting to see.

A midcourse correction on meditation and the brain: Richie's lab tried to repeat Sara Lazar's findings of cortical thickening by looking at long-term meditators, Westerners with day jobs and a minimum of five years as a practitioner—a group with an average lifetime 9,000 hours of meditation.[31] But the thickening Sara had reported did not show up, nor did several other structural changes that had been reported for MBSR.

There are many more questions than answers at this point. Some of the answers may come from data being analyzed as we write this, from Tania Singer's laboratory at the Max Planck Institute for Human Cognitive and Brain Sciences. There they are very carefully and systematically examining changes in cortical thickness associated with three different types of meditation practice (reviewed in chapter six, "Primed for Love"), in a massive study using a rigorous design with a large number of participants practicing over nine months.

One of the early findings to emerge from this work: different types of training are associated with different anatomical effects on the brain. For example, a method that emphasizes cognitive empathy and understanding how a person views life events was found to enhance cortical thickness in a specific region of the cortex toward the back of the brain, between the temporal and parietal lobes, known as the temporoparietal junction, or TPJ. In previous research by Tania's team, the TPJ has been found particularly active when we take another person's perspective.[32]

That brain change was found only with this method, and not with the others. Such findings underscore the importance for meditation researchers to distinguish among different types of practice, particularly when it comes to pinpointing related changes in the brain.

NEUROMYTHOLOGY

While we're spotlighting some of the neuromythology out and about concerning meditation, let's look at one bit that traces back to Richie's own research.[33] As of this writing, the best-known study from Richie's lab has 2,813 citations, an astonishing renown for an academic article. Dan was among the first to report on this research, in his book about the meeting in 2000 with the Dalai Lama on destructive emotions, where Richie presented this work in progress.[34]

The research has gone viral outside the academy, reverberating through the echo chamber of big and social media alike. And those bringing mindfulness to companies invariably mention it as "proof" the method will help folks there.

Yet that study raises large question marks in the eyes of scientists, especially Richie himself. We're talking about the time he had Jon Kabat-Zinn teach MBSR to volunteers at a high-stress biotech start-up where people were on the run virtually 24/7.

First, some background. For several years Richie pursued data on the ratio of activity in the right versus left prefrontal cortex while people were at rest. More right-side activity than left correlated with negative moods like depression and anxiety; relatively more left-side activity was associated with buoyant moods like energy and enthusiasm.

That ratio appeared to predict a person's day-to-day mood range. For the general population this ratio seemed to fit a bell curve, with most of us in the middle—we have good days and bad days. A very few people are at the extremes of the curve; if toward the left, they bounce back from feeling down, if toward the right they might be clinically anxious or depressed.

The study at the biotech start-up seemed to show a remarkable shift in brain function after the meditation training—from tilting toward the right to a leftward pitch, and reporting a switch into a more relaxed state. There were no such changes in a comparison group of workers assigned to a wait list, who were told they would receive the meditation training later.

But here's one major hitch. This research was never replicated, and was designed only as a pilot. We don't know, for instance, if an active control like HEP would result in similar benefits.

While that study was never replicated, others seem to support the finding on the brain ratio and its shift. A German study of patients with recurring episodes of severe depression found their ratio tilted strongly toward the right—which may be a neural marker of the disorder.[35] And the same German researchers found that this right-side tilt shifted back toward the left—but only while they were practicing mindfulness, not at normal rest.[36]

The problem: Richie's lab has not been able to show that this tilt toward left-side activation continues to grow stronger the more you meditate. Richie hit a snag when he started bringing to his lab

Olympic-level meditators, Tibetan yogis (more about them in chapter twelve, "Hidden Treasure"). These experts, who had logged off-the-charts hours of meditation, did not show the expected whopping leftward tilt—despite being some of the most optimistic and happy people Richie has ever known.

This undermined Richie's confidence in the measure, which he has discontinued. Richie has no sure sense of why that left/right measure failed to work as expected with the yogis. One possibility: a tilt toward the left may occur at the beginning of meditation practice, but other than a small range of change, the left/right ratio does not budge much. It may reflect temporary pressures or basic temperament but does not seem associated with enduring qualities of well-being or more complex changes in the brain found in those with high levels of meditation experience.

Our current thinking holds that in later stages of meditation other mechanisms kick in, so that what changes is your *relation* to any and all emotions, rather than the ratio of positive to negative ones. With high levels of meditation practice, emotions seem to lose their power to pull us into their melodrama.

Another possibility: different branches of meditation have disparate effects, so there may not be a clear line of development that's continuous from, say, mindfulness in beginners, to long-term vipassana practitioners, to the Tibetan experts assayed in Richie's lab.

And then there's the question of who teaches mindfulness. As Jon has told us, MBSR teachers vary greatly in expertise, in how much meditation retreat time they have put in, and in their own qualities of being. The biotech company had the benefit of having Jon himself as their teacher—over and above instruction in the MBSR techniques, he has unique gifts in imparting a view of reality that can potentially shift students' experience in ways that, possibly, might account for a shift in brain asymmetry. We don't know what the impacts would be if some other, randomly selected, MBSR teacher had come there.

THE BOTTOM LINE

Back to Dan and the meditation retreat he attended in hopes of lowering his blood pressure. Although he did get a big drop in his blood pressure readings immediately afterward, it's impossible to know whether it was because of the meditation or a more general "vacation effect," the relief we all feel when we drop our daily pressures and get away for a while.[37]

Within weeks his blood pressure readings were high again—and stayed that way until an astute physician guessed that he might have one of the few known causes of hypertension, a rare hereditary adrenal disorder. A medication that corrects that metabolic imbalance brought his blood pressure down to stay—while meditation did not.

Our questions are simple when it comes to whether meditation leads to better health: What's true, what's not, and what's not known? As we leapt into our survey of the hundreds of studies linking meditation to health effects, we applied strict standards. As is true of all too much meditation research, the methods used in many studies of health impacts fail to clear the highest bar. That left us surprised by how little we can say with certainty, given the great excitement (and, okay, hype) about meditation as a way to boost health.

The sounder studies, we found, focus on lessening our psychological distress rather than on curing medical syndromes or looking for underlying biological mechanisms. So, when it comes to a better quality of life for those with chronic diseases, yes to meditation. Such palliative care gets ignored too often in medicine, but it matters enormously to patients.

Still, might meditation offer biological relief? Consider the Dalai Lama, now in his eighties, who goes to bed at 7:00 p.m. and gets a full night's sleep before he awakens around 3:30 for a four-hour stint of spiritual practice, including meditation. Add another hour of practice before he goes to bed and that gives him five hours a day of contemplative time.

But painful arthritis in his knees makes going up or down stairs an ordeal—not uncommon for someone in the ninth decade of life. When he was asked if meditation helps medical conditions, he retorted, "If meditation was good for all health problems, I'd be free of pain in my knees."

When it comes to whether meditation does more than offer palliative help, we're not sure yet—and if so, in what medical conditions?

A few years after Richie got that stinging rejection of his plan to measure genetic changes from one day of meditation, he was invited to give the prestigious Stephen E. Straus Lecture at the National Institutes of Health, a yearly talk in honor of the founder of the National Center for Complementary and Integrative Health.[38]

Richie's topic, "Change Your Brain by Training Your Mind," was controversial, to say the least, among the many skeptics on the NIH campus. But, come the day of his talk, the august auditorium at the Clinical Center was packed, with many scientists watching a live stream from their offices—perhaps an augury of the changing status of meditation as a topic for serious research.

Richie's lecture focused on the findings in this area, mainly those from his lab, most of which are described in this book. Richie illuminated the neural, biological, and behavioral changes wrought by meditation, and how they might help maintain health—for instance, in better emotion regulation and sharpened attention. And, as we've tried to do here, Richie walked a very careful line between critical rigor and genuine conviction that there is really a "there" there: that meditation has beneficial impacts worthy of serious scientific investigation.

At the end of his talk, despite its staid academic tone, Richie received a standing ovation.

IN A NUTSHELL

None of the many forms of meditation studied here was originally designed to treat illness, at least as we recognize it in the West. Yet today the scientific literature is replete with studies assessing whether these ancient practices might be useful for treating just such illnesses. MBSR and similar methods can reduce the emotional component of suffering from disease, but not cure those maladies. Yet mindfulness training—even as short as three days—produces a short-term decrease in pro-inflammatory cytokines, the molecules responsible for inflammation. And the more you practice, the lower the level becomes of these pro-inflammatory cytokines. This seems to become a trait effect with extensive practice, with imaging studies finding in meditators at rest lower levels of pro-inflammatory cytokines, along with an increased connectivity between regulatory circuitry and sectors of the brain's self system, particularly the posterior cingulate cortex.

Among experienced meditation practitioners, a daylong period of intensive mindfulness practice down-regulates genes involved in inflammation. The enzyme telomerase, which slows cellular aging, increases after three months of intensive practice of mindfulness and loving-kindness. Finally, long-term meditation may lead to beneficial structural changes in the brain, though current evidence is inconclusive about whether such effects emerge with relatively short-term practice like MBSR, or only become apparent with longer-term practice. All in all, the hints of neural rewiring that undergird altered traits seem scientifically credible, though we await further studies for specifics.

10

Meditation as Psychotherapy

D r. Aaron Beck, the founder of cognitive therapy, had a question: "What is mindfulness?"

It was the mid-1980s, and Dr. Beck was asking Tara Bennett-Goleman, Dan's wife. She had come to his home in Ardmore, Pennsylvania, at Dr. Beck's request, because Judge Judith Beck, his wife, was about to undergo some elective surgery. Dr. Beck had a hunch meditation might help better prepare her mentally and, perhaps, even physically.

Tara instructed the couple on the spot. Following her guidance the Becks sat quietly and observed the sensations of their breathing in and out, then tried a walking meditation in their living room.

That was a hint of what has since become a strong movement in "mindfulness-based cognitive therapy," or MBCT. Tara's book *Emotional Alchemy: How the Mind Can Heal the Heart* was the first to integrate mindfulness with cognitive therapy.[1]

Tara had for years been a student of vipassana meditation and had recently completed a months-long intensive retreat with the Burmese meditation master U Pandita. That deep dive into the mind had yielded many insights, including one about the lightness of thoughts when viewed through the lens of mindfulness. That insight mirrors a principle in cognitive therapy of "decentering," observing thoughts and feelings without being overly identified with them. We can *reappraise* our suffering.

Dr. Beck had heard about Tara from one of his close students, Dr. Jeffrey Young, who at the time was establishing the first cognitive therapy center in New York City. Tara, with a freshly minted master's degree in counseling, was training with Dr. Young at his center. The two were jointly treating a young woman who suffered from panic attacks.

Dr. Young used a cognitive therapy approach, helping her distance herself from her catastrophizing thoughts—*I can't breathe, I'm going to die*—and challenge them. Tara brought mindfulness into the sessions, complementing Dr. Young's therapy approach with this unique lens on the mind. Learning to observe her breath mindfully—calmly and clearly, without panic—helped that patient overcome her panic attacks.

Working independently, psychologist John Teasdale at the University of Oxford, with Zindel Segal and Mark Williams, was writing *Mindfulness-Based Cognitive Therapy for Depression*, another such integration.[2] His research had revealed that for people with depression so severe that drugs or even electroshock treatments were no help, this mindfulness-based cognitive therapy (MBCT) cut the rate of relapse by half—more than any medication.

Such remarkable findings unleashed what has become a wave of research on MBCT. As has been true of most studies of meditation and psychotherapy, though, many of those studies (including Teasdale's original one) failed to meet the gold standard for clinical outcome research: randomized control groups and an equivalent comparison treatment by practitioners who believe theirs will bring results.

Some years later a group from Johns Hopkins University looked at what numbered by then forty-seven studies of meditation alone (that is, without including cognitive therapy) with patients suffering from distress ranging from depression and pain to sleep problems and overall quality of life—as well as maladies ranging from diabetes and arterial disease to tinnitus and irritable bowel syndrome.

This review, by the way, was exemplary in calculating the hours of meditation practice being studied: MBSR entailed twenty to twenty-seven hours of training over eight weeks, other mindfulness programs about half that. Transcendental meditation trials gave sixteen to thirty-nine hours over three to twelve months, and other mantra meditations about half that amount.

In a prominent article in one of the *JAMA* journals (the official publications of the American Medical Association), the researchers concluded that mindfulness (but not mantra-based meditation like TM, for which there were too few well-designed studies to make any conclusions) could lessen anxiety and depression, as well as pain. The degree of improvement was about as much as for medications, but without troubling side effects—making mindfulness-based therapies a viable alternative treatment for these conditions.

But no such benefits were found for other health indicators like eating habits, sleep, substance use, or weight problems. When it came to other psychological troubles, like ugly moods, addictions, and poor attention, the meta-analysis found little or no evidence that any kind of meditation might help—at least in the short-term interventions used in the research. Long-term meditation practice, they note, might well offer more benefits, though there were too little data on this for them to draw any conclusion.

The main problem: what had seemed promising for relieving problems from earlier studies of meditation disappeared into a mist when compared to the benefits from an active control like exercise. Bottom line for a wide range of stress-based problems: "insufficient evidence of any effect," at least as yet.[3]

From a medical perspective, these studies were the equivalent of a "low-dose, short-term" trial of a medication. The recommendation here: that more research be done, using far larger numbers of people and for a far longer period. That's quite apt for studies of treatments like a drug—the research model dominant in medicine. But such studies are enormously expensive, costing in the millions of dollars— and are paid for by drug companies or the National Institutes of Health. No such luck when it comes to meditation.

Another sticking point, and this a bit nerdy: the meta-analysis began by collecting 18,753 citations of articles of all kinds about meditation (a huge number, given that we could find but a paltry handful back in the 1970s, and just above 6,000 now—they used a broader number of search terms than we did). About half of those the authors spotted, though, were not reports of actual data; of the empirical reports, about 4,800 had no control group or were not randomized. After careful sifting, only 3 percent (that's the 47 in the analysis)—of the studies proved sufficiently well designed that they could be included in the review. As the Hopkins group points out, this simply underscores the need to upgrade meditation research.

This type of review carries great weight with physicians, in an era when medicine strives to become more evidence based. The Hopkins group did this meta-analysis for the Agency for Healthcare Research and Quality, whose guidelines physicians try to follow.

The review's conclusion: meditation (in particular, mindfulness) can have a role in treating depression, anxiety, and pain—about as much as medications but with no side effects. Meditation also can, to a lesser degree, reduce the toll of psychological stress. Overall, meditation has not been proven better for psychological distress than medical treatments, though the evidence for stronger conclusions remains insufficient.

But this was true as of 2013 (the study was published in January, 2014). With the quickened pace of meditation research, more and better-designed studies may overturn such judgments, at least to a degree.

Depression marks a singular case in point.

CHASING THE BLUES AWAY— WITH MINDFULNESS

The remarkable finding from John Teasdale's group at Oxford, that MBCT cut relapse in severe depression by around 50 percent, energized some impressive follow-up research. After all, a 50 percent drop in relapse outreaches by far what any medication used for severe depression can claim. If this beneficial impact were true of a drug, some pharmaceutical company would be minting money from it.

The need for more rigorous studies was clear; the original Teasdale pilot study had no control group, let alone a comparison activity. Mark Williams, one of Teasdale's original research partners at Oxford, spearheaded the research needed. His team recruited almost three hundred people with depression so severe that medications could not prevent them relapsing into doom and gloom—the same sort of difficult-to-treat patients as in the original study.

But this time the patients were randomly assigned to either MBCT or one of two active control groups where they either learned the basics of cognitive therapy or just had the usual psychiatric treatments.[4] The patients were tracked for six months to see if they had a relapse. MBCT proved more effective when it came to patients with a history of childhood trauma (which can make depression all the worse), and about the same as standard treatments with run-of-the-mill depression.

Soon after, a European group found that for a similar group with depression so severe that no medication helped them, MBCT did.[5] This, too, was a randomized study with an active control group. And by 2016 a meta-analysis of nine such studies with a total 1,258 patients concluded that, over a year afterward, MBCT was an effective way to lower the relapse rate in severe depression. The more severe the symptoms of depression, the larger the benefits from MBCT.[6]

Zindel Segal, one of John Teasdale's collaborators, delved more deeply into why MBCT seemed so effective.[7] He used fMRI to compare patients who had recovered from a bout of major depression, some of whom did MBCT, while the others received standard cognitive therapy (that is, without mindfulness). Those patients who, after treatments, showed a greater increase in the activity of their insula had 35 percent fewer relapses.

The reason? In a later analysis, Segal found the best outcomes were in those patients most able to "decenter," that is, step outside their thoughts and feelings enough to see them as just coming and going, rather than getting carried away by "*my* thoughts and feelings." In other words, these patients were more mindful. And the more time they put into mindfulness practice, the lower their odds of a relapse into depression.

At last a critical mass of research demonstrated to the satisfaction of the skeptical medical world that a mindfulness-based method could be effective for treating depression.

There are several variations of promising applications of MBCT for depression. For instance, women who are pregnant and have a previous history of depressive episodes naturally want to be sure they do not get depressed while carrying their baby or after the birth, and they are understandably leery of taking antidepressants while pregnant. Good news: a team led by Sona Dimidjian, another grad of the Summer Research Institute, found that MBCT could lower the depression risk in these women, and so offered a user-friendly alternative to drugs.[8]

When researchers from the Maharishi International University taught TM to prisoners with standard prison programs as the comparison, they found that four months later the prisoners doing TM showed fewer symptoms of trauma, anxiety, and depression; they also slept better and perceived their days as less stressful.[9]

Another instance: the angst-filled teen years can see the first onset of depressive symptoms. In 2015, 12.5 percent of the US population aged twelve to seventeen had at least one major depressive episode the previous year. This translates to about 3 million teens. While some of the more obvious signs of depression include negative thinking, severe self-criticism, and the like, sometimes the signs take subtle forms, like trouble sleeping or thinking or shortness of breath. A mindfulness program designed for teens reduced overt depression and such subtle signs, even six months after it ended.[10]

All of these studies, tantalizing as they are, need replication as well as upgrades to their design if they are to be acceptable to strict medical review standards. Still, for the person suffering from depressive bouts—or anxiety or pain—MBCT (and maybe TM) offers the possibility of relief.

Then there's the question of whether MBCT or meditation in alternate forms might relieve symptoms of other psychiatric maladies. And if so, what are the mechanisms that explain this?

Let's revisit that research on MBSR for people with social anxiety done by Philippe Goldin and James Gross at Stanford University (we reviewed it in chapter five). Social anxiety, which can look like anything from stage fright to shyness at gatherings, turns out to be a surprisingly common emotional problem, affecting more than 6 percent of the US population, around 15 million people.[11]

After the eight-week MBSR course the patients reported feeling less anxiety, a good sign. But you may recall the next step, which makes the study more intriguing: the patients also went into a brain scanner while doing a breath awareness meditation to manage their emotions as they listened to upsetting phrases like "people always judge me," one of the common fears in the mental self-talk among those with social anxiety. The patients reported feeling less anxious than usual on hearing such emotional triggers—and at the same time, brain activity lessened in their amygdala and increased in circuitry for attention.

This peek at the underlying brain activity may hint at the future of research on how meditation might relieve mental problems. For several years now—at least as of this writing—the National Institute of Mental Health (NIMH), the main source of funding for studies in this area, has disdained research that relies on the old categories of psychiatry listed in the field's *Diagnostic and Statistical Manual* (*DSM*).

While mental disorders like "depression" in its several varieties are in the *DSM*, the NIMH favors research that focuses on specific symptom clusters and their underlying brain circuitry—not just *DSM* categories. Along these lines, we wonder, for example, if the finding from Oxford, that MBCT works well with depressed patients who have a history of trauma, suggests that an overly reactive amygdala may be more involved in this treatment-resistant subgroup than among others who get depressed from time to time.

While we are pondering future research, here are a few more questions: What precisely is the added value of mindfulness compared with cognitive therapy? What disorders does meditation (including its use in MBSR and MBCT) relieve better than current standard psychiatric treatments? Should these methods be used along with those standard interventions? And what specific kinds of meditation work best to relieve which mental problems—and while we're at it, what's the underlying neural circuitry?

For now, these are unanswered questions. We're waiting to find out.

LOVING-KINDNESS MEDITATION
FOR TRAUMA

Recall that on September 11, 2001, a jet smashed into the Pentagon near Steve Z, and what had been an open office was instantly blasted into a haze-filled sea of wreckage, reeking of burned fuel. When the office was rebuilt he moved back to the very desk he had been sitting at on 9/11, but in a much lonelier setting—most of his office buddies had been killed in the fireball.

Steve recalls his feelings then: "We were fueled by rage: Those bastards—we'll get them! It was a dark place, a miserable time."

His severe PTSD was cumulative; Steve had previously served in combat theaters in Desert Storm and Iraq. The catastrophe of 9/11 intensified the trauma that had already been building.

For years after, anger, frustration, and hypervigilant distrust roiled within. But if anyone asked how he was doing, Steve's story line was, "No problem." He tried self-soothing with alcohol, hard jogging, visiting family, reading—anything to try to get a grip.

Steve was close to suicide when he entered Walter Reed Hospital for help, went through detox from alcohol, and slowly began the road to healing. He learned about his condition and agreed to meet with the psychotherapist he still sees, who introduced him to mindfulness meditation.

After two or three months of sobriety he tried to join a local mindfulness group, which met once a week. The first few times Steve went he walked in hesitantly, checked around the place, saw "these are not my people"—and walked out. Besides, he felt claustrophobic in closed spaces.

When he was finally able to try a short mindfulness retreat, he found it helped. And in particular what really clicked was the loving-kindness practice, a workable way to have compassion for himself as well as other people. With loving-kindness, he felt "at home again," a deep reminder of how he felt as a young boy playing with friends—a strong sense things were going to be okay.

"Practice helped me stay with those feelings and know, 'This will pass.' If I was getting angry, I could throw a little compassion and loving-kindness for myself and the other person."

Last we heard, Steve had gone back to school in mental health counseling, gotten credentialed as a psychotherapist, and was completing a clinical doctorate. His dissertation topic: "moral injury and spiritual wellness."

He connected with the Veterans Administration and support groups for military people like him with PTSD, and has been getting referrals from them for his small private practice. Steve feels uniquely equipped to help.

First findings say Steve's instincts had it right. At the Seattle Veterans Administration hospital, forty-two vets with PTSD took a twelve-week course in loving-kindness meditation, the kind Steve found helped him.[12] Three months later their PTSD symptoms had improved, and depression—a common side symptom—had lessened a bit.

These early findings are promising, but we don't know, say, if an active control condition like HEP would be just as effective. The caveats for the research on PTSD to date pretty much sum up the state of the art for scientific validation of meditation as a treatment for most psychiatric disorders.

Still, there are many arguments for compassion practice as an antidote to PTSD, beginning with anecdotal reports like Steve's.[13] Many are practical. A large proportion of veterans have PTSD; in any given year, between 11 and 20 percent of veterans suffer from PTSD, and over a veteran's lifetime that number goes up to 30 percent. If loving-kindness practice works, it offers a cost-effective group treatment.

Another reason: among the symptoms of PTSD are emotional numbness, alienation, and a sense of "deadness" in relationships—all of which loving-kindness might help reverse by the cultivation of positive feelings toward others. Still another: many vets dislike the side effects of the drugs they are given for PTSD, so they do not take them at all—and on their own are searching for nontraditional treatments. Loving-kindness appeals on both counts.

DARK NIGHTS

"I experienced a wave of self-hatred so shocking, so intense, that it changed the way I relate . . . to my own dharma path and the meaning of life itself." So recalls Jay Michaelson of the moment on a long, silent vipassana retreat when he fell into what he calls a "dark night" of intensely difficult mental states.[14]

The *Visuddhimagga* pegs this crisis as most likely at the point a meditator experiences the transitory lightness of thoughts. Right on schedule, Michaelson hit his dark night after having cruised through a quietly ecstatic landmark on that path, the stage of "arising and passing," where thoughts seem to disappear as soon as they begin, in rapid succession.

Shortly afterward he plunged into his dark night, a thick mixture of morbid doubt, self-loathing, anger, guilt, and anxiety. At one point the toxic mix was so strong, his practice collapsed; he broke down in tears.

But then he slowly began observing his mind rather than being sucked into the thoughts and feelings that swirled through it. He began to see these feelings as passing mental states, like any others. The episode was over.

Other such tales of meditative dark nights do not always have such a clean resolution; the meditator's suffering can be ongoing long after leaving the meditation center. Because the many positive impacts of meditation are far more widely known, some who go through dark nights discover people can't comprehend or even believe that they are hurting. All too frequently psychotherapists are little or no help.

Realizing the need, Willoughby Britton, a psychologist at Brown University (and a grad of the SRI), heads the "dark night project," which aids people who suffer from meditation-related psychological difficulties. Her Varieties of the Contemplative Experience project, as it is more formally called, adds to the more widely known beneficial impacts of meditation a caveat: When might it do harm?

At the moment, there are no firm answers. Britton has been collecting case studies and helping those who suffer from a dark night to understand what they are going through, that they are not alone, and, hopefully, to recover. Her study subjects have been largely referrals from guiding teachers at vipassana meditation centers where, over the years, there have been occasional dark night casualties during intensive retreats—despite those centers trying to weed out the vulnerable by asking on enrollment forms about psychiatric histories. To be sure, dark nights may not be related to such histories.

Dark nights are not unique to vipassana; most every meditative tradition warns about them. In Judaism, for example, Kabbalistic texts caution that contemplative methods are best reserved for middle age, lest an unformed ego fall apart.

At this point no one knows whether intensive meditation practice is in itself a danger to certain people, or if those who suffer dark nights might have had a breakdown of some sort no matter their circumstances. While Britton's case studies are anecdotal, their very existence is compelling.

The proportion of dark nights among all those who do prolonged retreats are, by all accounts, very small—though no one can say precisely what that proportion might be. From a research perspective, one of the findings needed would be to establish base rates for such difficulties both among meditators and in the population at large.

Nearly one in five adults in the United States, nearly 44 million, were found by the National Institute of Mental Health to suffer from a mental illness in any given year. Both freshman year at college and military boot camp—and even psychotherapy—are known to precipitate psychological crises in a certain small percentage of people. The research question becomes, Is there something about deep meditation that puts some people at risk over and above this base rate?

For those who do have such a dark night, Willoughby Britton's program offers practical advice and comfort. And despite the (rather low) risk of dark nights, especially during prolonged retreats, meditation has come into vogue among psychotherapists.

MEDITATION AS METATHERAPY

In Dan's first article on meditation he proposed it might be used in psychotherapy.[15] That article, "Meditation as Meta-Therapy," appeared during Dan's 1971 sojourn in India, and nary a psychotherapist showed much interest. Yet on his return he somehow was invited to lecture on this notion at a meeting of the Massachusetts Psychological Association.

After his talk ended, a slim, bright-eyed young man wearing an ill-fitting sport jacket approached him, saying he was a graduate student in psychology with similar interests. He had spent several years as a monk in Thailand studying meditation, surviving there on the generosity of the Thai people, a country where every household finds it an honor to feed monks. No such luck in New England.

This grad student thought that as a psychologist he could adapt meditative tools, in the guise of psychotherapy, to alleviate people's suffering. He was glad to hear someone else was making the connection between meditation and therapeutic applications.

That grad student was Jack Kornfield, on whose dissertation committee Richie served. Jack became one of the founders of, first, the Insight Meditation Society in Barre, Massachusetts, and then went on to found Spirit Rock, a meditation center in the San Francisco Bay Area. Jack has been a pioneer in translating Buddhist theories of the mind into language attuned to the modern sensibility.[16]

Jack, along with a group including Joseph Goldstein, designed and ran the teacher training program that graduated the very teachers who helped Steve Z recover from his PTSD all those years later. Jack's own explanation of Buddhist psychological theories, *The Wise Heart*, shows how this perspective on the mind and working with meditation can be used in psychotherapy—or on your own. This synthesis was the first of his by now many books integrating traditional Eastern and modern approaches.

Another main voice in this movement has been Mark Epstein, a psychiatrist. Mark was a student in Dan's psychology of consciousness course, and, as a Harvard senior, he asked Dan to be his faculty adviser for an honors project on Buddhist psychology. Dan, at the time the only member of the Harvard psychology department with interest and a bit of knowledge in the area, agreed; Mark and Dan later wrote an article together in a short-lived journal.[17]

In a series of books integrating psychoanalytic and Buddhist views of mind, Mark has continued to lead the way. His first book had the intriguing title *Thoughts Without a Thinker*, a phrase from the object relations theorist Donald Winnicott, which also voices a contemplative perspective.[18] Tara's, Mark's, and Jack's works are emblematic of a wider movement, with countless therapists now blending various contemplative practices or perspectives with their own approach to psychotherapy.

While the research establishment remains somewhat skeptical of the potency of meditation as a treatment for DSM-level disorders, the widening pool of psychotherapists enthusiastic about bringing together meditation and psychotherapy continues to grow. Although researchers await randomized studies with active controls, psychotherapists already offer meditation-enriched treatments for their clients.

For instance, as of this writing there have been 1,125 articles in the scientific literature on mindfulness-based cognitive therapy. Tellingly, more than 80 percent of these were published in the past five years.

Of course, meditation has its limits. Dan's original interest in meditation during his college days was because he felt anxious. Meditation seemed to calm those feelings somewhat, but they still came and went.

Many people go to psychotherapists for just such problems. Dan did not. But years later he was diagnosed with that adrenal disorder, the cause of his long-standing high blood pressure. One of those adrenal symptoms: elevated levels of cortisol, the stress hormone that triggers feelings of anxiety. Along with his years of meditation, a drug that adjusts that adrenal problem seemed also to handle the cortisol—and the anxiety.

IN A NUTSHELL

Although meditation was not originally intended to treat psychological problems, in modern times it has shown promise in the treatment of some, particularly depression and anxiety disorders. In a meta-analysis of forty-seven studies on the application of meditation methods to treat patients with mental health problems, the findings show that meditation can lead to decreases in depression (particularly severe depression), anxiety, and pain—about as much as medications but with no side effects. Meditation also can, to a lesser degree, reduce the toll of psychological stress. Loving-kindness meditation may be particularly helpful to patients suffering from trauma, especially those with PTSD.

The melding of mindfulness with cognitive therapy, or MBCT, has become the most empirically well-validated psychological treatment with a meditation basis. This integration continues to have a wide impact in the clinical world, with empirical tests of applications to an ever larger range of psychological disorders under way. While there are occasional reports of negative effects of meditation, the findings to date underscore the potential promise of meditation-based strategies, and the enormous increase in scientific research in these areas bodes well for the future.

A Yogi's Brain

11

n the steep hills above the ridge-hugging Himalayan village of McLeod Ganj, you might stumble on a small hut or remote cave housing a Tibetan yogi on a long-term, solo retreat. In the spring of 1992, an intrepid team of scientists, Richie and Cliff Saron among them, traveled to those huts and caves to assess the brain activity of the yogi within each.

A three-day journey had brought them to McLeod Ganj, the hill station in the foothills of the Himalayas that is home to the Dalai Lama and the Tibetan Government-in-Exile. There the scientists set up shop in a guesthouse owned by a brother of the Dalai Lama, who resides nearby. Several rooms were given over to unpacking and assembling the equipment for deployment in backpacks for transport up to the mountain hermitages.

In those days such brain measurements required a mélange of

EEG electrodes and amplifiers, computer monitors, video recording equipment, batteries, and generators. That equipment, much larger than today's, weighed several hundred pounds. Traveling with those instruments in their hard protective cases, the researchers resembled a nerdy rock band. There were no roads to follow; yogis on retreat choose the most remote place they can find. And so, with great effort, and the help of several porters, the scientists lugged their measuring instruments to the yogis.

The Dalai Lama himself had identified these yogis as masters in *lojong*, a systematic mind training method; in his view these were ideal subjects for study. The Dalai Lama had written a letter urging the yogis to cooperate, and even sent along a personal emissary, a monk from his private office, to vouch for the top-level request that they participate.

Arriving at a yogi's hermitage, the scientists presented the letter and through a translator asked to monitor the yogi's brain while he meditated.

The same answer came from each yogi in turn: No.

To be sure, they all were exceptionally friendly and warm. Some offered to teach the scientists the very practices they wanted to measure. A few said they would think about it. But none would go ahead then and there.

Some may have heard about another yogi who once had been persuaded by a similar letter from the Dalai Lama to leave his retreat and travel to a university in faraway America to demonstrate his ability to raise his core body temperature at will. That yogi had died soon after his return, and rumors on the mountainside held that the experiment had played a role.

For most of these yogis, science was quite foreign; none had much inkling of the role of science in modern Western culture. Moreover, of the eight yogis the team met on this expedition, only one had ever seen an actual computer before Richie and the team arrived.

A few of the yogis made the canny argument that they had no idea what, exactly, the strange machines measured. If the measurements were irrelevant to what they were doing, or if their brain failed to meet some scientific expectation, it might look to some as though their methods were of no use. That, they said, might discourage those on the same path.

Whatever the reasons, the net result of this scientific expedition was a resounding nothing.

Despite the failure to get cooperation, let alone data, and though futile in the short term, the exercise proved instructive, beginning a steep learning curve. For starters, better to bring the meditators to the equipment, especially in a well-fortified brain lab—if they would come.

For another, research on such adepts confronts unique challenges beyond their rarity, their intentional remoteness, and their unfamiliarity with or disinterest in scientific endeavors. While their mastery at this inner expertise seems akin to world-class rankings in sports, in this "sport," the better you get the less you care about your ranking—let alone social status, riches, or fame.

That list of indifferences includes any personal pride you might take in what scientific measures show about your inner accomplishments. What mattered to them was how the results might influence others for better or worse.

Prospects for scientific studies were dim.

A SCIENTIST AND A MONK

Enter Matthieu Ricard, whose degree in molecular genetics from France's Pasteur Institute had been under the tutelage of François Jacob, who later won the Nobel Prize in Medicine.[1] As a postdoc Matthieu abandoned his promising career in biology to become a monk; over the decades since, he has lived in retreat centers, monasteries, and hermitages.

Matthieu was an old friend of ours; he had often participated (as had we) in dialogues (organized by the Mind and Life Institute) between the Dalai Lama and various groups of scientists, where Matthieu voiced the Buddhist viewpoint on whatever topic was at hand.[2] You might recall that during the "destructive emotions" dialogue, the Dalai Lama exhorted Richie to test meditation rigorously and extract what was of value for the benefit of the larger world.

The Dalai Lama's call to action touched Matthieu as strongly as it did Richie, stirring in this monk's mind (to his surprise) a long-unused expertise in the scientific method. Matthieu himself was the first monk to come for study at Richie's lab, spending several days as experimental subject and as collaborator on methods to refine the protocol used with a succession of other yogis. Matthieu Ricard was a coauthor on the main journal article reporting initial findings with yogis.[3]

For much of the time Matthieu had spent as a monk in Nepal and Bhutan, he was the personal attendant to Dilgo Khyentse Rinpoche, one of the last century's most universally revered Tibetan meditation masters.[4] Many, many lamas of note among those living in exile from Tibet—including the Dalai Lama—had sought out Dilgo Khyentse for private instruction.

This put Matthieu at the heart of a large network within the Tibetan meditative world. He knew whom to suggest as potential subjects of study—and, perhaps most important, was trusted by those very meditation experts. Matthieu's participation made all the difference in recruiting those elusive adepts.

Matthieu could reassure them that there was good reason to travel half the globe to the university campus in Madison, Wisconsin—a place many Tibetan lamas and yogis had never heard of, let alone seen. Further, they would have to put up with the weird food and habits of a foreign culture.

To be sure, some of those recruited had taught in the West and were familiar with its cultural norms. But, beyond the journey to an exotic land, there were the strange rituals of the scientists—in the yogis' eyes an entirely alien endeavor. For those more familiar with Himalayan hermitages than with the modern world, nothing in their frame of reference made much sense of all this.

Matthieu's reassurance that their efforts would be worthwhile was the key to their cooperation. For these yogis, "worthwhile" did not mean their participation would have a personal benefit—increase their fame or feed their pride—but rather that it would help other people. As Matthieu understood, their motivation was compassion, not self-interest.

Matthieu emphasized the motivation of the scientists, who dedicated themselves to this because they believed if the scientific evidence supported the efficacy of these practices, it would help promote the incorporation of the practices into Western culture.

Matthieu's crucial reassurances have so far brought twenty-one of these most advanced meditators to Richie's lab for brain studies. That number includes seven Westerners who have done at least one three-year retreat at the center in Dordogne, France, where Matthieu has practiced, as well as fourteen Tibetan adepts who traveled to Wisconsin from India or Nepal.

FIRST, SECOND, AND THIRD PERSONS

Matthieu's training in molecular biology gave him an ease with the rigors and rules of science's methods. He dove into the planning sessions to help design the methods that would be used to assay the first guinea pig—himself. As both design collaborator and volunteer number one, he tried out the very scientific protocol he had helped shape.

While extremely unusual in the annals of science, there are precedents for researchers to be the first guinea pig in their own experiments, particularly to be assured of the safety of some new medical treatment.[5] The rationale here, though, was not fear of exposing others to an unknown risk, but rather, a unique consideration when it comes to studying how we might train the mind and shape the brain.

What's being studied is intensely private, one person's inner experience—while the tools used to measure it are machines that yield objective measures of biological reality, but nothing of that inner one. Technically, the inner assessment requires a "first-person" report, while the measurements are a "third-person" report.

Closing the gap between the first and third person was the idea of Francisco Varela, the brilliant biologist and cofounder of the Mind and Life Institute. In his academic writing Varela proposed a method for combining the first- and third-person lenses with a "second person," an expert on the topic being studied.[6] And, he argued, the person being studied should have a well-trained mind, and so, yield better data than someone not so well trained.

Matthieu was both topic expert and possessor of that well-trained mind. So, for example, when Richie began to study the various types of meditation, he did not realize that "visualization" required more than just generating a mental image. Matthieu explained to Richie and his team that the meditator also cultivates a particular emotional state that goes along with a given image—say, with an image of the bodhisattva Tara the accompanying state melds compassion and loving-kindness. Advice such as this led Richie's group to change from being guided by the top-down norms of brain science, to collaborating with Matthieu in the details of designing the experimental protocol. [7]

Long before Matthieu became a collaborator we had moved in this direction by immersing ourselves in what we were studying—meditation—to generate hypotheses for empiric testing. These days science knows this general approach as an instance of the generation of "grounded theory"—that is, grounded in a direct personal sense of what's going on.

Varela's approach goes a step further, necessary when what's being studied lurks in the mind and brain of one person yet resembles a strange land to the one doing the research. Having experts like Matthieu in this private domain allows methodological precision where there would otherwise be guesswork.

We admit to our own mistakes here. Back in the 1980s, when Richie was a young professor at the State University of New York at Purchase and Dan a journalist working in New York City, we joined together for some research on a single, gifted meditator. This student of U Ba Khin (Goenka's teacher) had himself become a teacher, and claimed he could enter at will a state of *nibbana*—the endpoint of that Burmese meditative path. We wanted to find hard correlates of that vaunted state.

Problem was, the main tool we had was an assay of blood levels of cortisol, a hot topic in research of the day. We used that as our main measure because we were borrowing a lab from one of the main investigators of cortisol—not because there was some strong hypothesis relating *nibbana* to cortisol. But taking cortisol levels demanded that the meditator—ensconced in a hospital room on the other side of a one-way mirror—be hooked up to an IV that let us draw his blood every hour; we traded shifts with two other scientists so we could provide around-the-clock coverage, a routine we followed for several days.

The meditator signaled with a buzzer several times during those few days that he had entered *nibbana*. But the cortisol levels budged not at all—they were irrelevant. We also deployed a brain measure, but that, too, was not so apt, and primitive by today's standards. We've come a long way.

What might be next as contemplative science continues to evolve? The Dalai Lama, a twinkle in his eye, once told Dan that someday "the person being studied and the person doing the research will be one and the same."

Perhaps partly with that aim in mind, the Dalai Lama has encouraged a group at Emory University to introduce a Tibetan-language science curriculum into the studies of monks in monasteries.[8] A radical move: the first such change in six hundred years!

THE JOY OF LIVING

One cool September morning in 2002, a Tibetan monk arrived at the Madison, Wisconsin, airport. His journey had started seven thousand miles away, at a monastery atop a hill on the fringe of Kathmandu,

Nepal. The trip took eighteen hours in the air over three days and crossed ten time zones.

Richie had met the monk briefly at the 1995 Mind and Life meeting on destructive emotions in Dharamsala, but had forgotten what he looked like. Still, it was easy to pick him out from the crowd. He was the only shaven-headed man wearing gold-and-crimson robes in the Dane County Regional Airport. His name was Mingyur Rinpoche and he had traveled all this way to have his brain assayed while he meditated.

After a night's rest, Richie brought Mingyur to the lab's EEG room, where brain waves are measured with what looks like a surrealist art piece: a shower cap extruding spaghetti-like wires. This specially designed cap holds 256 thin wires in place, each leading to a sensor pasted to a precise location on the scalp. A tight connection between the sensor and the scalp makes all the difference between recording usable data about the brain's electrical activity and having the electrode simply be an antenna for noise.

As Mingyur was told when a lab technician began pasting sensors to his scalp, ensuring a tight connection for each and placing them in their exact spot takes no more than fifteen minutes. But when Mingyur, a shaven-headed monk, offered up his bald scalp, it turned out such continually exposed skin is more thickened and callused than one protected by hair. To make the crucial electrode-to-scalp connection tight enough to yield viable readings through thicker skin ended up taking much longer than usual.

Most people who come into the lab get impatient, if not irritated, by such delays. But Mingyur was not in the least perturbed, which calmed the nervous lab technician—and all those looking on—with the feeling that anything that happened would be okay with him.

That was the first inkling of Mingyur's ease of being, a palpable sense of relaxed readiness for whatever life might bring. The lasting impression Mingyur conveyed was of endless patience and a gentle quality of kindness.

After spending what seemed like an eternity ensuring that the sensors had good contact with the scalp, the experiment was finally ready to begin. Mingyur was the first yogi studied after that initial session with Matthieu. The team huddled in the control room, eager to see if there was a "there" there.

A precise analysis of something as squishy as, say, compassion demands an exacting protocol, one that can detect that mental state's specific pattern of brain activity amid the cacophony of the electrical storm from everything else going on. The protocol had Mingyur alternate between one minute of meditation on compassion and thirty seconds of a neutral resting period. To ensure confidence that any effect detected was reliable rather than a random finding, he would have to do this four times in rapid succession.

From the start Richie had grave doubts about whether this could work. Those on the lab team who meditated, Richie among them, all knew it takes time just to settle the mind, often considerably longer than a few minutes. It was inconceivable, they thought, that even someone like Mingyur would be able to enter these states instantaneously and not need some time to reach inner quiet.

Despite their skepticism, in designing this protocol they had listened to Matthieu, who knew both the culture of science and of the hermitage. He had assured them that these mental gymnastics would be no problem for someone at Mingyur's level of expertise. But Mingyur was the first such adept to be formally tested this way and Richie and his technicians were unsure, even nervous.

Richie was fortunate that John Dunne, a Buddhist scholar at the University of Wisconsin—who exhibits a rare combination of scientific interests, humanities expertise, and fluency in Tibetan—volunteered to translate.[9] John delivered precisely timed instructions to Mingyur signaling him to start a compassion meditation, and then after sixty seconds another cue for thirty seconds of his mental resting state, and so on for three more cycles.

Just as Mingyur began the meditation, there was a sudden huge burst of electrical activity on the computer monitors displaying the signals from his brain. Everyone assumed this meant he had moved; such movement artifacts are a common problem in research with EEG, which registers as wave pattern readings of electrical activity at the top of the brain. Any motion that tugs the sensors—a leg shifting, a tilt of the head—gets amplified in those readings into a huge spike that looks like a brain wave and has to be filtered out for a clean analysis.

Oddly, this burst seemed to last the entire period of the compassion meditation, and as far as anyone could see, Mingyur had not moved an iota. What's more, the giant spikes diminished but did not disappear as he went into the mental rest period, again with no visible shift in his body.

The four experimenters in the control room team watched, transfixed, while the next meditation period was announced. As John Dunne translated the next instruction to meditate into Tibetan, the team studied the monitors in silence, glancing back and forth from the brain wave monitor to the video trained on Mingyur.

Instantly the same dramatic burst of electrical signal occurred. Again Mingyur was perfectly still, with no visible change in his body's position from resting to the meditation period. Yet the monitor still displayed that same brain wave surge. As this pattern repeated each time he was instructed to generate compassion, the team looked at one another in astonished silence, nearly jumping off their seats in excitement.

The lab team knew at that moment they were witnessing something profound, something that had never before been observed in the laboratory. None could predict what this would lead to, but everyone sensed this was a critical inflection point in neuroscience history.

The news of that session has created a scientific stir. As of this writing, the journal article reporting these findings has been cited more than 1,100 times in the world's scientific literature.[10] Science has paid attention.

A MISSED BOAT

About the time news of Mingyur Rinpoche's remarkable data was reaching the scientific world, he was invited to the lab of a famous cognitive scientist then at Harvard University. There Mingyur was put through two protocols: in one he was asked to generate an elaborate visual image; in the other he was assessed to see if he had any knack for extrasensory perception. The cognitive scientist had high hopes that he would document the achievements of an extraordinary subject.

Mingyur's translator, meanwhile, was fuming because the protocol was not just hours long and onerous but painfully irrelevant to Mingyur's actual meditative expertise—from the translator's perspective, an act of disrespect within Tibetan norms for treating a teacher like Mingyur (who nevertheless retained his usual good cheer throughout).

The net result of Mingyur's day in that lab: he flunked both tests, doing no better than the college sophomores who were the usual subjects of study there.

Mingyur, it turned out, had done no practice with visualization since the long-gone, early years of his practice. As time went on, his meditations evolved. His current method, ongoing open presence (which expresses itself as kindness in everyday life), encourages letting go of any and all thoughts rather than generating any specific visual images. Mingyur's practice actually ran counter to the purposeful generation of an image and the feelings that go along with it—perhaps reversing any skill he might once have had in that. His circuitry for visual memory had gotten no particular workout, despite his thousands of hours spent in other kinds of mental training.

As for "extrasensory perception," Mingyur had never claimed to have such supernormal powers. Indeed, the texts of his tradition made clear that any fascination with such abilities was a detour, a dead end on the path.

That was no secret. But nobody had asked him. Mingyur had run smack into a paradox of today's research on consciousness, the mind, and meditative training: those who do the research on meditation are too often in the dark about what they are actually studying.

Ordinarily in the cognitive neurosciences, a "subject" (the term for someone who volunteers for the study, in the objectifying, at-a-distance language of science) goes through an experimental protocol designed by the researcher. The researcher concocts that design without conferring with any of the subjects, partly because subjects are meant to be naive about the purpose (to avoid a potential biasing factor) but also because the scientists have their own points of reference—their hypotheses, other studies done in the field which they hope to inform, and the like. Scientists don't consider their subjects particularly well informed about any of this.

That traditional scientific stance completely missed the chance to assess Mingyur's actual meditative talents, as did our earlier failure to take the measure of *nibbana*. Both times that first- and third-person estrangement led to misjudging where these meditators' remarkable strengths lie and how to measure them, akin to testing a legendary golfer like Jack Nicklaus on his prowess at shooting basketball free throws.

NEURAL PROWESS

Back to Mingyur's time in Richie's lab. The next stunner came when Mingyur went through another batch of tests, this time with fMRI, which renders what amounts to a 3-D video of brain activity. The fMRI gives science a lens that complements the EEG, which tracks the brain's electrical activity. The EEG readings are more precise in time, the fMRI more accurate in neural locations.

An EEG does not reveal what's happening deeper in the brain, let alone show *where* in the brain the changes occur—that spatial precision comes from the fMRI, which maps the regions where brain activity occurs in minute detail. On the other hand, fMRI, though spatially exacting, tracks the timing of changes over one or two seconds, far slower than EEG.

While his brain was probed by the fMRI, Mingyur followed the cue to engage compassion. Once again the minds of Richie and the others watching in the control room felt as though they had stopped. The reason: Mingyur's brain's circuitry for empathy (which typically fires a bit during this mental exercise) rose to an activity level 700 to 800 percent greater than it had been during the rest period just before.

Such an extreme increase befuddles science; the intensity with which those states were activated in Mingyur's brain exceeds any we have seen in studies of "normal" people. The closest resemblance is in epileptic seizures, but those episodes last brief seconds, not a full minute. And besides, brains are *seized* by seizures, in contrast to Mingyur's display of intentional control of his brain activity.

Mingyur was a meditation prodigy, as the lab team learned while tallying his history of lifetime hours of the practice: at that point, 62,000. Mingyur grew up in a family of meditation experts; his brother Tsoknyi Rinpoche and half brothers Chokyi Nyima Rinpoche and Tsikey Chokling Rinpoche are considered contemplative masters in their own right.

Their father, Tulku Urgyen Rinpoche, was widely respected among the Tibetan community as one of the few great living masters in this inner art who had trained in old Tibet, but then (spurred by China's invasion) lived outside that country. While Mingyur has as of this writing been on retreats for a total of ten of his forty-two years, Tulku Urgyen reputedly had done more than twenty years of retreat over his lifetime; Mingyur's grandfather—Tulku Urgyen's father—was said to have put in more than thirty years on retreat.[11]

As a young boy one of Mingyur's favorite pastimes was pretending he was a yogi meditating in a cave. He entered a three-year meditation retreat when he was just thirteen, a decade or more earlier than most who undertake such a challenge. And by the end of that retreat he proved so proficient that he was made meditation master for the next three-year round, which began soon after the first ended.

THE WANDERER RETURNS

In June 2016, Mingyur Rinpoche came back to Richie's lab. It had been eight years since Mingyur had last been studied there. We were fascinated to see what an MRI of his brain might show.

Some years before, he had announced he would be starting another three-year retreat—his third. But to everyone's shock, instead of going into a remote hermitage with an attendant along to cook and care for him as is traditional, he disappeared one night from his monastery in Bodh Gaya, India, taking only his robes, a bit of cash, and an ID card.

During his odyssey Mingyur lived as a wandering mendicant, spending winters as a sadhu on the plains of India and during the warmer months inhabiting Himalayan caves where fabled Tibetan masters had stayed. Such a wandering retreat, not uncommon in old Tibet, has become rare, especially among Tibetans like Mingyur whose diaspora has brought them into the modern world.

During those wandering years there was not a word from him, save once when he was recognized by a Taiwanese nun at a retreat cave. He gave her a letter (telling her to send it after he had moved on) that said not to worry, he was fine—and exhorting his students to practice. A photo that surfaced when a monk, a longtime friend, managed to join Mingyur shows a radiant face with a wispy beard and long hair, his expression one of ebullient rapture.

Then, suddenly, in November 2015, after almost four and a half years as a wanderer in radio silence, Mingyur reappeared at his monastery in Bodh Gaya. On hearing that news, Richie arranged to see him during a visit to India that December.

Months later, Mingyur stopped in Madison while on an American teaching tour, and stayed at Richie's house. Within minutes of his arrival at the house Mingyur agreed to go back into the scanner. Only a few months after returning from his hardscrabble life he seemed right at home in this up-to-the-minute lab.

As Mingyur entered the MRI suite, the lab technician gave him a friendly welcome, saying, "I was the tech the last time you were in the scanner." Mingyur beamed his electric smile in return. While he waited for the machine to be readied, Mingyur joked with another member of Richie's team, an Indian scientist from Hyderabad.

Given the go-ahead, Mingyur left his sandals at the bottom of the two-step ladder that boosted him to the MRI table and lay down so the tech could strap his head into a cradle tight enough that it allows no more than 2 millimeters of movement—all the better to obtain sharp images of his brain. His calves, thickened by years of trekking the steep slopes of the Himalayas, protruded from his monk's robes and then disappeared as the table slipped into the maw of the MRI.

The technology had improved since his last visit; the monitors reveal a crisper image of his brain's folds and tucks. It would take months to compare these data with those collected years before, and to track the changes in his brain during that time against the normal alterations seen in the brains of men his age.

Although he was barraged with requests, following his return from this last retreat, to have his brain scanned by many labs all over the world, Mingyur turned most all of them down for fear of becoming a perpetual subject. He had consented to have his brain rescanned by Richie and his team because he knew they had longitudinal data from previous scans and could analyze ways his brain might be showing atypical changes.

The first scan Richie's lab had of Mingyur's brain was obtained in 2002; there was another in 2010 and now the most recent, in 2016. These three scans provided the lab team with an opportunity to examine age-related declines in gray matter density, the site of the brain's molecular machinery. Each of us has a decrease in the density of gray matter as we age, and as we saw in chapter nine, "Mind, Body, and Genome," a given brain can be compared with a large database of the brains of other people the same age.

With the development of high-resolution MRI, scientists have now discovered that they can use anatomical landmarks to estimate the age of a person's brain. Brains of people of a given age group into a normal distribution, a bell-shape curve; most people's brain's hover around their chronological age. But some people's brains age more quickly than their chronological age would predict, putting them at risk for premature age-related brain disorders such as dementia. And other people's brains age more slowly compared with their chronological age.

As of this writing the most recent set of scans of Mingyur's brain are still being processed, but Richie and his team see some clear patterns already, using rigorous quantitative anatomical landmarks. Comparing Mingyur's brain to norms for his age, he falls in the 99th percentile—that is, if we had 100 people who are the same chronological age as Mingyur (forty-one years at this scan), his brain would be the youngest in a group of 100 age- and gender-matched peers. After his latest retreat as a wanderer, when the lab compared Mingyur's brain changes to those of a control group over the same period of time, Mingyur's brain is clearly aging more slowly.

Although his chronological age was forty-one at the time, his brain fit most closely the norms for those whose chronological age was thirty-three.

This rather remarkable fact highlights the further reaches of neuroplasticity, the very basis of an altered trait: an enduring mode of being reflecting an underlying change in the structure of the brain.

The total hours of practice Mingyur put in during his years as a wanderer are difficult to calculate. At his level of expertise, "meditation" becomes an ongoing feature of awareness—a trait—not a discrete act. In a very real sense, he practices continuously, day and night. In fact, in his lineage the distinction made is not the conventional equation of meditation with time spent in a session sitting on a cushion versus regular life, but rather, between being in a meditative state or not, no matter what else you are doing.

From Mingyur's very first visit to the lab, he had rendered compelling data hinting at the power of intentional, sustained mental exercise to redesign neural circuitry. But the findings from Mingyur were only anecdotal, a single case that might be explained many different ways. For instance, perhaps his remarkable family has some mysterious genetic predisposition that both motivates them to meditate and leads them to high levels of proficiency.

More convincing are results from a larger group of seasoned meditation adepts like Mingyur. His remarkable neural performance was part of a larger story, a one-of-a-kind brain research program that has harvested data from these world-class meditation experts. Richie's lab continues to study and analyze the mass of data points from these yogis, in an ever-growing set of findings unparalleled in the history of contemplative traditions, let alone brain science.

IN A NUTSHELL

At first Richie's lab found it impossible to get the cooperation of the most highly experienced yogis. But when Matthieu Ricard, a seasoned yogi himself with a PhD in biology, assured his peers their participation might be of benefit to people, a total of twenty-one yogis agreed. Matthieu, in an innovative collaboration with Richie's lab, helped design the experimental protocol. The next yogi to come to the lab, Mingyur Rinpoche, was also the one with most lifetime hours of practice—62,000 at the time. When he meditated on compassion there was a huge surge in electrical activity in his brain as recorded by EEG; fMRI images revealed that during this meditation his circuitry for empathy jumped in activity by 700 to 800 percent compared to its level at rest. And when he later went on retreat as a wanderer for four and a half years, the aging of his brain slowed, so that at forty-one his brain resembled the norm for thirty-three-year-olds.

Hidden Treasure

While Mingyur's visit to Madison had yielded jaw-dropping results, he was not alone. Over the years in Richie's lab, those twenty-one yogis have come to be formally tested. They were at the height of this inner art, having racked up lifetime meditation hours ranging from 12,000 to Mingyur's 62,000 (the number he had accomplished while going through these studies, and before his four-years-plus wandering retreat).

Each of these yogis completed at least one three-year retreat, during which they meditated in formal practice a minimum of eight hours per day for three continuous years—actually, for three years, three months, and three days. That equates, in a conservative estimate, to about 9,500 hours per retreat.

All have undergone the same scientific protocol, those four one-minute cycles of three kinds of meditation—which has yielded a mountain of metrics. The lab's team spent months and months analyzing the dramatic changes they saw during those short minutes in these highly seasoned practitioners.

Like Mingyur, they entered the specified meditative states at will, each one marked by a distinctive neural signature. As with Mingyur, these adepts have shown remarkable mental dexterity, instantly and with striking ease mobilizing these states: generating feelings of compassion, the spacious equanimity of complete openness to whatever occurs, or laser-sharp, unbreakable focus.

They entered and left these difficult-to-achieve levels of awareness within split seconds. These shifts in awareness were accompanied by equally pronounced shifts in measurable brain activity. Such a feat of collective mental gymnastics has never been seen by science before.

A SCIENTIFIC SURPRISE

Recall that at the last minute the bedridden Francisco, just a month before he died, had to cancel attending that meeting in Madison with the Dalai Lama. He sent his close student Antoine Lutz, who had just received his PhD under Francisco's mentorship, to present in his absence.

Richie and Antoine met for the first time just one day before that meeting, and from the start their two scientific minds melded. Antoine's background in engineering and Richie's in psychology and neuroscience made for a complementary pairing.

Antoine ended up spending the next ten years in Richie's lab, where he brought his precision mind to the analysis of the EEGs and fMRIs of yogis. Antoine, like Francisco, has been a dedicated meditation practitioner himself, and the combination of his introspective insights with his scientific mind-set made for an extraordinary colleague in Richie's center.

Now a professor at the Lyon Neuroscience Research Center in France, Antoine continues to pursue research in contemplative neuroscience. He has been involved from the start in the research with yogis and has coauthored a stream of articles, with more coming, reporting his findings.

Preparing the raw data on the yogis for sifting by sophisticated statistical programs has demanded painstaking work. Just teasing out the differences between the yogis' resting state and their brain activity during meditation was a gargantuan computing task. So it took Antoine and Richie quite a while to stumble upon a pattern hiding in that data flood, empirical evidence that got lost amid the excitement about the yogis' prowess in altering their brain activity during meditative states. In fact, the missed pattern surfaced only as an afterthought during a less hectic moment, months later, when the analytic team sifted through the data again.

All along the statistical team had focused on temporary state effects by computing the difference between a yogi's baseline brain activity and that produced during the one-minute meditation periods. Richie was reviewing the numbers with Antoine and wanted a routine check to ensure that the initial baseline EEG readings—those taken at rest, before the experiment began—were the same in a group of control volunteers who tried the identical meditations the yogis were doing. He asked to see just the baseline measures by themselves.

When Richie and Antoine sat down to review what the computers had just crunched, they looked at the numbers and then looked at one another. They knew exactly what they were seeing and exchanged just one word: "Amazing!"

All the yogis had elevated gamma oscillations, not just during the meditation practice periods for open presence and compassion but also during the very first measurement, before any meditation was performed. This electrifying pattern was in the EEG frequency known as "high-amplitude" gamma, the strongest, most intense form. These waves lasted the full minute of the baseline measurement before they started the meditation.

This was the very EEG wave that Mingyur had displayed in that surprising surge during both open presence and compassion. And now Richie's team saw that same unusual brain pattern in all the yogis as a standard feature of their everyday neural activity. In other words, Richie and Antoine had stumbled upon the holy grail: a neural signature showing an enduring transformation.

There are four main types of EEG waves, classed by their frequency (technically, measured in hertz). Delta, the slowest wave, oscillates between one and four cycles per second, and occurs mainly during deep sleep; theta, the next slowest, can signify drowsiness; alpha occurs when we are doing little thinking and indicates relaxation; and beta, the fastest, accompanies thinking, alertness, or concentration.

Gamma, the very fastest brain wave, occurs during moments when differing brain regions fire in harmony, like moments of insight when different elements of a mental puzzle "click" together. To get a sense of this "click," try this: What single word can turn each of these into a compound word: *sauce, pine, crab?**

The instant your mind comes up with the answer, your brain signal momentarily produces that distinctive gamma flare. You also elicit

---

*Answer: apple.

a short-lived gamma wave when, for instance, you imagine biting into a ripe, juicy peach and your brain draws together memories stored in different regions of the occipital, temporal, somatosensory, insular, and olfactory cortices to suddenly mesh the sight, smells, taste, feel, and sound into a single experience. For that quick moment the gamma waves from each of these cortical regions oscillate in perfect synchrony. Ordinarily gamma waves from, say, a creative insight, last no longer than a fifth of a second—not the full minute seen in the yogis.

Anyone's EEG will show distinctive gamma waves for short moments from time to time. Ordinarily, during a waking state we exhibit a mixture of different brain waves that wax and wane at different frequencies. These brain oscillations reflect complex mental activity, like information processing, and their various frequencies correspond to broadly different functions. The location of these oscillations varies among brain regions; we can display alpha in one cortical location and gamma in another.

In the yogis, gamma oscillations are a far more prominent feature of their brain activity than in other people. Our usual gamma waves are not nearly as strong as that seen by Richie's team in yogis like Mingyur. The contrast between the yogis and controls in the intensity of gamma was immense: on average the yogis had twenty-five times greater amplitude gamma oscillations during baseline compared with the control group.

We can only make conjectures about what state of consciousness this reflects: yogis like Mingyur seem to experience an ongoing state of open, rich awareness during their daily lives, not just when they meditate. The yogis themselves have described it as a spaciousness and vastness in their experience, as if all their senses were wide open to the full, rich panorama of experience.

Or, as a fourteenth-century Tibetan text describes it,

> ... *a state of bare, transparent awareness;*
> *Effortless and brilliantly vivid, a state of relaxed, rootless*
>   *wisdom;*
> *Fixation free and crystal clear, a state without the slightest*
>   *reference point;*
> *Spacious empty clarity, a state wide-open and unconfined;*
>   *the senses unfettered* ...[1]

The gamma brain state Richie and Antoine discovered was more than unusual, it was unprecedented—a *wow!* No brain lab had ever before seen gamma oscillations that persist for minutes rather than split seconds, are so strong, and are in synchrony across widespread regions of the brain.

Astonishingly, this sustained, brain-entraining gamma pattern goes on even while seasoned meditators are asleep—as was found by the Davidson group in other research with long-term vipassana meditators who have an average of about 10,000 hours lifetime practice. These gamma oscillations continuing during deep sleep are, again, something never seen before and seem to reflect a residual quality of awareness that persists day and night.[2]

The yogis' pattern of gamma oscillation contrasts with how, ordinarily, these waves occur only briefly, and in an isolated neural location. The adepts had a sharply heightened level of gamma waves oscillating in synchrony across their brain, independent of any particular mental act. Unheard of.

Richie and Antoine were seeing for the first time a neural echo of the enduring transformations that years of meditation practice etch on the brain. Here was the treasure, hidden in the data all along: a genuine altered trait.

STATE BY TRAIT

In one of the many studies Antoine spearheaded, when volunteers new to meditation were trained for a week in the same practices that the yogis do, there was absolutely no difference between the volunteers' brains at rest and when they were trying to meditate on cue, as the yogis did.[3] This contrasts with the remarkable difference between resting and meditation in the yogis. Since any learnable mental skill takes sustained practice over time to master, given the massive hours of lifetime meditation among the yogis, we are not surprised by this vast difference between novices and masters.

But there's another surprise here: the yogis' remarkable talent at entering a specific meditative state on cue, within a second or two, itself signals an altered trait. This mental feat stands in stark contrast to most of us meditators who, relative to the yogis, are more like beginners: when we meditate, it takes us a while to settle our minds, let go of distracting thoughts that overwhelm our focus, and get some momentum in our meditation.

From time to time we may have what we consider a "good" meditative experience. And now and then we might peek at our watch to see how much longer the session should last.

Not for the yogis.

Their remarkable meditation skills bespeak what's technically known as a "state by trait interaction," suggesting the brain changes that underlie the trait also give rise to special abilities that activate during meditative states—here, a heightened speed of onset, greater intensity, and extended duration.

In contemplative science, an "altered state" refers to changes that occur only during meditation. An altered trait indicates that the practice of meditation transformed the brain and biology so that meditation-induced changes are seen *before* beginning to meditate.

So a "state-by-trait" effect refers to temporary state changes that are seen only in those who display enduring altered traits—the long-term meditators and the yogis. Several have shown up during the research in Richie's lab.

One example. Recall that the yogis show a pronounced elevation in gamma activity during the open presence and compassion meditations, far greater than in the controls. This elevation in gamma activity was a change from baseline, their everyday levels—marking another state-by-trait effect.

What's more, while they rest in "open presence," the very distinction between a state and a trait blurs: in their tradition, the yogis are explicitly instructed to mingle the state of open presence with their everyday life—to morph the state into a trait.

READY FOR ACTION

One by one they lay in the scanners, their heads held firmly in place by cumbersome earphones. There was one group of meditation novices, and another of Tibetan and Western yogis (lifetime average 34,000 hours); each one had his or her (yes, there were female yogis) brain scanned while doing a compassion practice.[4]

The specific method they deployed was described by Matthieu

Ricard, who collaborated on the study, as follows. First bring to mind someone you care about deeply and relish the feeling of compassion toward that person—and then hold that same loving-kindness toward all beings, without thinking of anyone in particular.[5]

During the session of loving-kindness each person heard at random a series of sounds, some happy, like a baby laughing; others neutral, like background sounds in a café, or still others, sounds of human suffering (like screams, as in the studies in chapter six). Just as in previous studies of empathy and the brain, for everyone the neural circuitry for tuning in to distress activated more strongly during compassion meditation than when those vocal signals of suffering came while the person was at rest.

Significantly, this brain response for sharing another person's feelings was greater in the yogis compared to beginners. In addition, their expertise in compassion practice also upped action in circuitry typically involved while we sense another person's mental state or take their perspective. Finally, there was a boost in brain areas, especially the amygdala, key for what's salient; we feel another person's distress is of compelling importance and pay more attention.

Tellingly, the yogis but not the beginners showed the final part of the brain's arc to action, a jump in activity in the motor centers that guide the body when we are ready to move—to take some decisive action to help, even though the subjects were lying still in a scanner. The yogis showed a huge boost in these circuits. The involvement of neural regions for action, particularly the premotor cortex, seems striking: to emotional resonance with a person's suffering it adds the readiness to help.

The yogi's neural profile during compassion seems to reflect an endpoint of the path of change. For people who have never meditated before, absolute beginners, the pattern does not show up during their meditation on compassion—it takes a bit of practice. There's a dose response here: this pattern shows up a bit in beginners, more in people who have put in more lifetime hours of meditation, and to the greatest extent in the yogis.

Intriguingly, yogis hearing sounds of people in distress while they were doing loving-kindness meditation showed less activity than others do in their postcingulate cortex (PCC), a key area for self-focused thought.[6] In the yogis, hearing sounds of suffering seems to prime a focus on others.

They also show a stronger connection between the PCC and the prefrontal cortex, an overall pattern suggesting a "down-regulation" of the "what will happen to me?" self-concern that can dampen compassionate action.[7]

Some of the yogis later explained that their training imbued them with *preparedness* for action, so the moment they encounter suffering they are predisposed to act without hesitation to help the person. This preparedness, along with their willingness to engage with someone's suffering, counters the normal tendency to withdraw, to back away from a person in distress.

That seems to embody the advice of Tibetan meditation master (and Matthieu's main teacher) Dilgo Khyentse Rinpoche to yogis such as these: "Develop a complete acceptance and openness to all situations and emotions, and to all people, experiencing everything totally without mental reservations and blockages. . . ."[8]

PRESENCE TO PAIN

An eighteenth-century Tibetan text urges meditators to practice "on whatever harms come your way," adding, "When sick, practice on that sickness. . . . When cold, practice on that coldness. By practicing in this way all situations will arise as meditation."⁹

Mingyur Rinpoche, likewise, encourages making all sensation, even pain, our "friend," using it as a basis for meditation. Since the essence of meditation is awareness, any sensation that anchors attention can be used as support—and pain particularly can be very effective in focusing. Treating it as a friend "softens and warms" our relationship, as he puts it, as we gradually learn to accept the pain rather than try to get rid of it.

With that advice in mind, consider what happened when Richie's group used the thermal stimulator to create intense pain in the yogis. Each yogi (including Mingyur) was compared to a meditation-naive volunteer matched for age and gender. For a week before they came to be studied, these volunteers learned to generate an "open presence," an attentional stance of letting whatever life presents us come and go, without adding thoughts or emotional reactions. Our senses are fully open, and we just stay aware of what happens without getting carried away by any downs or ups.

All those in the study were first tested to find their individual maximal heat point. Then they were told they would get a ten-second blast of that fiery device, which would be preceded by a slight warming of the plate—a ten-second warning. Meanwhile, their brain was being scanned.

The moment the plate heated a bit—the cue for pain about to come—the control groups activated regions throughout the brain's pain matrix as though they were already feeling the intense burn. The reaction to the "as if" pain—technically, "anticipatory anxiety"—was so strong that when the actual burning sensation began, their pain matrix activation became just a bit stronger. And in the ten-second recovery period, right after the heat subsided, that matrix stayed nearly as active—there was no immediate recovery.

This sequence of anticipation-reactivity-recovery gives us a window on emotion regulation. For instance, intense worry about something like an upcoming painful medical procedure can in itself cause us anticipatory suffering, just imagining how bad we will feel. And after the real event we can continue to be upset by what we have gone through. In this sense our pain response can start well before and last long after the actual painful moment—exactly the pattern shown by those volunteers in the comparison group.

The yogis, on the other hand, had a very different response in this sequence. They, like the controls, were also in a state of open presence—no doubt one some magnitudes greater than for the novices. For the yogis, their pain matrix showed little change in activity when the plate warmed a bit, even though this cue meant extreme pain was ten seconds away. Their brains seemed to simply register that cue with no particular reaction.

But during the actual moments of intense heat the yogis had a surprising heightened response, mainly in the sensory areas that receive the granular feel of a stimulus—the tingling, pressure, high heat, and other raw sensations on the skin of the wrist where the hot plate rested. The emotional regions of the pain matrix activated a bit, but not as much as the sensory circuitry.

This suggests a lessening of the psychological component—like the worry we feel in anticipation of pain—along with intensification of the pain sensations themselves. Right after the heat stopped, all the regions of the pain matrix rapidly returned down to their levels before the pain cue, far more quickly than was the case for the controls. For these highly advanced meditators, the recovery from pain was almost as though nothing much had happened at all.

This inverted V-shaped pattern, with little reaction during

anticipation of a painful event, followed by a surge of intensity at the actual moment, then swift recovery from it, can be highly adaptive. This lets us be fully responsive to a challenge as it happens, without letting our emotional reactions interfere before or afterward, when they are no longer useful. This seems an optimal pattern of emotion regulation.

Remember the fear we felt when we were six years old about going to the dentist to get a cavity filled? This could mean nightmares at that age. But we change as we grow older. When we are twenty-six, what might have loomed as a trauma in childhood becomes ho-hum, an appointment to schedule in the midst of a busy day. We are a very different person as an adult than we were as a child—we bring more mature ways of thinking and reacting to the moment.

Likewise, with the yogis in the pain study, their many years of meditation practice suggests the state they were in during the pain reflects enduring changes acquired through their training. And because they were engaged in the open presence practice, this, too, qualifies as a state by trait effect.

EFFORTLESS

As with any skill we sharpen, within the first weeks of meditation practice, beginners notice increased ease. For instance, when volunteers new to meditation practiced daily for ten weeks, they reported the practice progressively got easier and more enjoyable, whether they were focusing on their breath, generating loving-kindness, or just observing the flow of their thoughts.[10]

And as we saw in chapter eight, Judson Brewer found a group of long-term meditators (with an average lifetime practice of about 10,000 hours) reported effortless awareness during meditation in association with decreased activity in the PCC, that part of the default network active during "selfing" mental operations.[11] When we take the self out of the picture, it seems, things go along with little effort.

When long-term meditators reported "undistracted awareness," "effortless doing," "not efforting," and "contentment," activation in the PCC went down. On the other hand, when they reported "distracted awareness," "efforting," and "discontentment," activation of the PCC went up.[12]

A group of first-time meditators also reported an increase in ease, though only while they were actively being mindful—a state effect that did not persist otherwise. For the beginners, "increased ease" appears very relative: going from exerting great effort—particularly to counter the mind's tendency to wander—and getting a bit better at it as the days and weeks go on. But the easing of their effort goes nowhere near the effortlessness found in the yogis, as we've seen in their remarkable performance in the on/off lab protocol.

One metric for effortlessness here comes down to being able to keep your mind on a chosen point of focus and resist the natural tendency to wander off into some train of thought or be pulled away by a sound, while having no feeling of making an effort. This kind of ease seems to increase with practice.

Richie's lab group initially compared expert meditators to controls in the magnitude of prefrontal activation during focused attention on a small light. The long-term meditators showed a modest increase in prefrontal activation compared with the controls, though the difference was, strangely, not very impressive.

One afternoon as Richie and his lab team sat around a long

conference table pondering these somewhat disappointing data, they began to reflect on the large span of expertise even within the so-called expert meditator group. This expert group actually ranged in practice hours from 10,000 to 50,000—a very large spread. Richie wondered what they would find if they compared those with the most versus least amount of practice. He had already found that with higher levels of expertise, there's an effortlessness that actually would be reflected in *less* rather than *more* prefrontal activation.

When the team compared those with the most versus those with the least amount of practice, they found something truly striking: all of the increase in prefrontal activation was accounted for by those with the *least* amount of practice. For those with the most lifetime hours of practice, there was very little prefrontal activation.

Curiously, the activation tended to occur only at the very beginning of a practice period, while the mind was focusing on the object of concentration, that little light. Once the light was in focus, the prefrontal activation dropped away. This sequence may represent the neural echoes of effortless concentration.

Another measure of concentration was to see how distracted the meditators are by emotional sounds—laughing, screaming, crying—which they heard in the background while focusing on the light. The more amygdala activation in response to those sounds, the more wavering in concentration. Among meditators with the greatest amount of lifetime practice hours—an average of 44,000 lifetime hours (the equivalent of twelve hours a day for ten years) the amygdala hardly responded to the emotional sounds. But for those with less practice, (though still a high number—19,000 hours) the amygdala also showed a robust response. There was a staggering 400 percent difference in the size of the amygdala response between these groups!

This indicates an extraordinary selectivity of attention: a brain effortlessly able to block out the extraneous sounds and the emotional reactivity they normally elicit.

What's more, this means traits continue to alter even at the highest level of practice. The dose-response relationship does not seem to end even up to 50,000 hours of practice.

The finding of a switch to effortlessness in brain function among the most highly experienced yogis was only possible because Richie's group had assessed total lifetime hours of meditation practice. Lacking that simple metric, this valuable finding would have been buried in the general comparison between novices and experts.

THE HEART-MIND

Back in 1992, Richie and that gallant band of researchers brought their tons of equipment to India, hoping to measure the most seasoned meditation masters near where the Dalai Lama lives. Next to his residence sits the Namgyal Monastery Institute of Buddhist Studies, an important training ground for monk-scholars in the Dalai Lama's tradition. Richie and his researcher friends, you'll remember, were unable to collect any real scientific data from the mountain-dwelling yogis.

But when the Dalai Lama asked Richie and his colleagues to give a talk on their work to the monks in the monastery, Richie thought maybe the equipment they schlepped to India could be put to some good use. Rather than just a dry academic talk, they would give a live demonstration of how brain electrical signals can be recorded.

And so, two hundred monks were dutifully sitting on cushions on the floor when Richie and friends arrived with their suitcases filled with EEG equipment. To place a headful of electrodes takes quite a bit of time. Richie and the other scientists worked as quickly as possible to secure all the electrodes in place.

The demo that evening used as subject the neuroscientist Francisco Varela. As Richie placed the electrodes on Francisco's scalp, the view of Francisco was blocked. But when Richie completed his task and moved out of way, a loud chorus of laughter erupted from the usually very staid monks.

Richie thought the monks were laughing because Francisco looked a bit funny with wires coming off his scalp electrodes like a big bundle of spaghetti. But that was not what the monks found funny.

They were laughing because Richie and his team had told them of their interest in studying compassion—but they were placing electrodes on the head, rather than the heart!

It took Richie's group about fifteen years to see the monks' point. Once yogis started to come to Richie's lab, the group saw data that made them realize compassion was very much an embodied state, with tight links between the brain and body, and especially between the brain and the heart.

Evidence for this linkage came from an analysis that related the yogis' brain activity to their heart rate—a follow-up to the unexpected finding that the yogis' hearts beat more rapidly compared to novices' when they heard sounds of people in distress.[13] The yogis' heart rate was coupled with the activity of a key area in the insula, a brain region that acts as the portal through which information about the body is conveyed to the brain and vice versa.

In a sense, then, the Namgyal monks were right. Richie's team had data suggesting that with yogic training the brain becomes more finely tuned to the heart—specifically during compassion meditation.

Again, this was a state-by-trait finding, one that occurred in the yogis only when they meditated on compassion (and not during other kinds of meditation, at rest, or among those in a comparison group).

In short, compassion in the yogis sharpens their sense of other people's emotions, especially if they are distraught, and heightens sensitivity to their own bodies—particularly the heart, a key source of empathic resonance with the suffering of others.

The variety of compassion may matter. Here the practitioners were engaged in "nonreferential" compassion. In the words of Matthieu, they were "generating a state in which love and compassion permeated the whole mind with no other discursive thoughts." They were not focusing on any specific person, but rather were generating the background quality of compassion; this may be especially important in engaging the neural circuits that tune the brain to the heart.

Being present to another person—a sustained, caring attention—can be seen as a basic form of compassion. Careful attention to another person also enhances empathy, letting us catch more of the fleeting facial expressions and other such cues that attune us to how that person actually feels in the moment. But if our attention "blinks," we may miss those signals. As we saw in chapter seven, long-term meditators have fewer such blinks in their attention than do other folks.

This cancellation of the attentional blink numbers among a host of mental functions that change with rigorous mind training—and which scientists had thought to be frozen, immutable, basic properties of the nervous system. Most of these are little known outside scientific circles, where they are taken as strong givens—a challenge to that status jars the assumptive system of cognitive science. But discarding old assumptions in light of new findings is the motor of science itself.

Another point. We expect that the lightening of self and lessening of attachment in the yogis would correlate with a shrinking of the nucleus accumbens, as was found in long-term Western meditators. But Richie has collected no data on this from the yogis, despite the falling away of attachments being an explicit goal of their practice.

The discovery of the default mode and how to measure it, as well as its crucial role in the brain's self-system, has come so recently that when the yogis were coming one by one through the lab, Richie's team had no inkling they might want to use the baseline to measure this shift. Only toward the tail end of this stream did the lab get the resting state measures needed—and on too few yogis to have robust data for the analysis.

Science progresses in part through innovative measures that yield data never seen before. That's what we have here. But that also means the slices of findings we have on the yogis have more to do with the serendipity of measures available to us than with some careful assay of the topography of this region of human experience.

This highlights a weakness in what otherwise might seem quite impressive findings on the yogis: these data points are but glimpses of the altered traits that intensive, prolonged meditation produces. We do not want to reduce this quality of being to what we happen to be able to measure.

Science's view of these yogis' altered traits is akin to the parable of the blind men and the elephant. The gamma finding, for instance, seems quite exciting, but it's like feeling the elephant's trunk without knowing about the rest of its body. And so, too, with their missing attentional blink, effortless meditative states, ultrarapid recovery from pain, and readiness to help someone in distress—these are but glimpses of a larger reality we do not fully comprehend.

What matters most, though, may be the realization that our ordinary state of waking consciousness—as William James observed more than a century ago—is but one option. Altered traits are another.

A word about the global significance of these yogis. Such people are very rare, what some Asian cultures call "living treasures." Encounters with them are extremely nourishing and often inspiring, not because of some vaunted status or celebrity but because of the inner qualities they radiate. We hope nations and cultures that harbor such beings will see the need to protect them and their communities of expertise and practice, as well as preserve the cultural attitudes that value these altered traits. To lose the way to this inner expertise would be a world tragedy.

IN A NUTSHELL

The massive levels of gamma activity in the yogis and the synchrony of the gamma oscillations across widespread regions of the brain suggest the vastness and panoramic quality of awareness that they report. The yogis' awareness in the present moment—without getting stuck in the anticipation of the future or ruminating on the past—seems reflected in the strong "inverted V" response to pain, where yogis show little anticipatory response and very rapid recovery. The yogis also show neural evidence of effortless concentration: it takes only a flicker of the neural circuitry to place their attention on a chosen object, and little to no effort to hold it there. Finally, when generating compassion, the brains of yogis become more connected to their bodies, particularly their hearts—indicating emotional resonance.

Altering Traits

I n the beginning nothing comes, in the middle nothing stays, in the end nothing goes." That enigmatic riddle comes from Jetsun Milarepa, Tibet's eminent twelfth-century poet, yogi, and sage.[1]

Matthieu Ricard unpacks Milarepa's puzzle this way: at the start of contemplative practice, little or nothing seems to change in us. After continued practice, we notice some changes in our way of being, but they come and go. Finally, as practice stabilizes, the changes are constant and enduring, with no fluctuation. They are altered traits.

Taken as a whole, the data on meditation track a rough vector of progressive transformations, from beginners through the long-term meditators and on to the yogis. This arc of improvement seems to reflect both lifetime hours of practice as well as time on retreat with expert guidance.

The studies of *beginners* typically look at the impacts from under 100 total hours of practice—and as few as 7. The *long-term* group, mainly vipassana meditators, had a mean of 9,000 lifetime hours (the range ran from 1,000 to 10,000 hours and more).

And the *yogis* studied in Richie's lab, had all done at least one Tibetan-style three-year retreat, with lifetime hours up to Mingyur's 62,000. Yogis, on average had three times more lifetime hours than did long-term meditators—9,000 hours versus 27,000.

A few long-term vipassana meditators had accumulated more than 20,000 lifetime hours and one or two up to 30,000, though none had done a three-year retreat, which became a de facto distinguishing feature of the yogi group. Despite the rare overlaps in lifetime hours, the vast majority of the three groups fall into these rough categories.

There are no hard-and-fast lifetime hour cutoffs for the three levels, but research on them has clustered in particular ranges. We've organized meditation's benefits into three dose-response levels, roughly mapping on the novice to amateur to professional rankings found in expertise of all kinds, from ballerinas to chess champions.

The vast majority of meditators in the West fall into the first level: people who meditate for a short period—a few minutes to half an hour or so on most days. A smaller group continues on to the long-term meditator level. And a mere handful attain the expertise of the yogis.

Let's look at the impacts in those who have just begun a meditation practice. When it comes to stress recovery, the evidence for some benefits in the first few months of daily practice are more subjective than objective—and shaky. On the other hand the amygdala, a key node in the brain's stress circuitry, shows lessened reactivity after thirty or so hours over eight weeks of MBSR practice.

Compassion meditation shows stronger benefits from the get-go; as few as seven total hours over the course of two weeks leads to increased connectivity in circuits important for empathy and positive feelings, strong enough to show up outside the meditation state per se. This is the first sign of a state morphing into a trait, though these effects likely will not last without daily practice. But the fact that they appear outside the formal meditation state itself may reflect our innate wiring for basic goodness.

Beginners also find improvements in attention very early on, including less mind-wandering after *just eight minutes* of mindfulness practice— a short-lived benefit, to be sure. But even as little as two weeks of practice is sufficient to produce less mind-wandering and better focus and working memory, enough for a significant boost in scores on the GRE, the entrance exam for graduate school. Indeed, some findings suggest decreases in activation in the self-relevant regions of the default mode with as little as two months of practice. When it comes to physical health, there is more good news: small improvements in the molecular markers of cellular aging seem to emerge with just thirty hours of practice.

Still, all such effects are unlikely to persist without sustained practice. Even so, these benefits strike us as surprisingly strong for beginners. Take-home: practicing meditation can pay off quickly in some ways, even if you have just started.

IN THE LONG TERM

Sticking with meditation over the years offers more benefits as meditators reach the long-term range of lifetime hours, around 1,000 to 10,000 hours. This might mean a daily meditation session, and perhaps annual retreats with further instruction lasting a week or so—all sustained over many years. The earlier effects deepen, while others emerge.

For example, in this range we see the emergence of neural and hormonal indicators of lessened stress reactivity. In addition, functional connectivity in the brain in a circuit important for emotion regulation is strengthened, and cortisol, a key hormone secreted by the adrenal gland in response to stress, lessens.

Loving-kindness and compassion practice over the long term enhance neural resonance with another person's suffering, along with concern and a greater likelihood of actually helping. Attention, too, strengthens in many aspects with long-term practice: selective attention sharpens, the attentional blink diminishes, sustained attention becomes easier, and an alert readiness to respond increases. And long-term practitioners show enhanced ability to down-regulate the mind-wandering and self-obsessed thoughts of the default mode, as well as weakening connectivity within those circuits—signifying less self-preoccupation. These improvements often show up during meditative states, and generally tend to become traits.

Shifts in very basic biological processes, such as a slower breath rate, occur only after several thousand hours of practice. Some of these impacts seem more strongly enhanced by intensive practice on retreat than by daily practice.

While evidence remains inconclusive, neuroplasticity from long-term practice seems to create both structural and functional brain changes, such as greater working connection between the amygdala and the regulatory circuits in the prefrontal areas. And the neural circuits of the nucleus accumbens associated with "wanting" or attachment appear to shrink in size with longer-term practice.

While in general we see a gradient of shifts with more lifetime meditation hours, we suspect there are different rates of change in disparate neural systems. For instance, the benefits of compassion come sooner than does stress mastery. We expect studies in the future will fill in the details of a dose-response dynamic for various brain circuits.

Intriguing signs suggest that long-term meditators to some degree undergo state-by-trait effects that enhance the potency of their practice. Some elements of the meditative state, like gamma waves, may continue during sleep. And a daylong retreat by seasoned meditators benefited their immune response at the genetic level—a finding that startled the medical establishment.

THE YOGIS

At this world-class level (roughly 12,000 to 62,000 lifetime hours of practice, including many years in deep retreat), truly remarkable effects emerge. Practice in part revolves around converting meditative states to traits—the Tibetan term for this translates as "getting familiar" with the meditative mind-set. Meditation states merge with daily activities, as altered states stabilize into altered traits and become enduring characteristics.

Here Richie's group saw signs of altered traits in the yogi's brain function and even structure, along with strongly positive human qualities. The jump in synchronized gamma oscillations initially observed during compassion meditation was also found, albeit to a lesser extent, in the baseline state. In other words, for the yogis this state has become a trait.

State-by-trait interactions mean that what happens during meditation can be very different for the yogis, showing up starkly when compared with novices doing the same practice. Perhaps the strongest evidence comes from the yogis' response to physical pain during simple mindfulness-type practice: a sharp "inverted V," with little brain activity during anticipation of the pain, an intense but very short peak during the pain, followed by very rapid recovery.

For most of us who meditate, concentration takes mental effort, but for the yogis with most lifetime hours, it becomes effortless. Once their attention locks onto a target stimulus, their neural circuits for effortful attention go quiet while their attention stays perfectly focused.

When the yogis meditate on compassion there's a strengthening of the coupling between heart and brain beyond what is ordinarily seen. Finally, there is that tantalizing bit of data showing shrinking in the nucleus accumbens in long-term meditators, suggesting we might find further structural changes in the yogi's brain that support a lessening of attachment, grasping, and self-focus. Precisely what other such neural shifts there might be, and what they mean, await deciphering in future research.

THE AFTER

These remarkable data points merely hint at the full flowering of the contemplative path at this level. Some of these findings have shown up through happenstance—as when Richie decided to check on the baseline data for the yogis, or to look at the most seasoned group compared to the rest.

And then there's anecdotal evidence: when Richie's lab asked one yogi to take swabs of saliva to assess his cortisol activity while he was on retreat, the levels were so low they were off the standard scale, and the lab had to adjust the assay range downward.

Some Buddhist traditions speak of this level of stabilization as recognition of an inner "basic goodness" that permeates the person's mind and activities. As one Tibetan lama said about his own teacher—a master revered by all the Tibetan contemplative lineages—"Someone like him has a two-tier consciousness," where his meditative accomplishments are a steady background for whatever else he does.

Several labs—including Richie's and Judson Brewer's—have noticed that more advanced meditators can show a brain pattern while merely resting that resembles that of a meditative state like mindfulness or loving-kindness, while beginners do not.[2] That comparison of the expert meditator's baseline with someone new to practice stands as a hallmark of the way altered traits show up in research, though it offers just a snapshot.

Perhaps one day an ultralong study will give us the equivalent of a video on how altered traits emerge. For now, as the Brewer group conjectured, meditation seems to transform the resting state—the brain's default mode—to resemble the meditative state.

Or, as we put it long ago, the after is the before for the next during.

IN SEARCH OF LASTING CHANGE

"If the heart wanders or is distracted," advised Francis de Sales (1567–1622), a Catholic saint, "bring it back to the point quite gently . . . and even if you did nothing during the whole of your hour but bring your heart back . . . though it went away every time, your hour would be very well-employed."[3]

Virtually all meditators execute a common series of steps, no matter the specifics of practice. These begin with an intended focus—but then after a while your mind wanders off. When you notice it has wandered you can make the final step: bring your mind back to the original focus.

Research at Emory University by Wendy Hasenkamp (an SRI alum and now director of science at Mind and Life Institute) found the connections between brain regions involved in these steps to be stronger among more seasoned meditators.[4] Importantly, the differences between meditator and controls were found not just in meditation but in the ordinary "resting" state as well—suggesting a possible trait effect.

The lifetime hours measure offers a ripe opportunity to correlate that number with, say, changes in the brain. But to be sure such an association is not due to self-selection or other such factors requires another step: a longitudinal study where, ideally, the impact grows stronger as practice continues (plus an active control group followed for the same length of time who do not show those changes).

Two longitudinal studies—Tania Singer's work on empathy and compassion, and Cliff Saron's on shamatha—have yielded some of the most convincing data yet on the power of meditation to create altered traits. And then there are some surprises.

Take a finding from Tania's research. She notes that some researchers had wondered why meditators who did a daily body scan (as in Goenka's method) failed to show any improvement in counting their heartbeats, a standard test of "interoception," or attunement to the body.

She found an answer in her ReSource Project. The ability to be aware of bodily signals like heartbeat did not increase after three months of daily practice of "presence," which includes a mindful body scan. However, those very improvements began to show up after six months, with even bigger gains after nine months. Some benefits take time to ripen—what psychologists call a "sleeper" effect.

Consider the tale of a yogi who had spent years in a Himalayan cave on retreat. One day a traveler happened by and, seeing the yogi, asked him what he was doing. "I'm meditating on patience," the yogi said.

"In that case," replied the traveler, "you can go to hell!"

To which the yogi angrily retorted, "You go to hell!"

That tale (like the one about the yogi in the bazaar) has served for centuries as a cautionary tale to serious practitioners, reminding them that the test of their practice is life itself, not isolated hours in meditation. A trait like patience should leave us unflappable no matter what life brings our way.

The Dalai Lama told this story, clarifying, "There's a saying in Tibetan that in some cases practitioners have the outward form of being holy people, which holds when everything is fine—when the sun is shining and the belly is full. But when they are confronted with a real challenge or crisis, then they become just like everyone else."[5]

The "full catastrophe" of our lives offers the best durability test of altered traits. While a yogi's superlow cortisol level on retreat tells us how relaxed he can get, his cortisol level during a hectic day would reveal whether this had become a permanent, altered trait.

EXPERTISE

We've all heard it takes someone 10,000 hours of practice to master a skill like computer programming or golf, right?

Wrong.

In reality science finds that some domains (like memorization) can be mastered in as little as 200 hours. More to the point, Richie's lab finds that even among the meditation adepts—all of whom have put in at least 10,000 hours of practice—expertise continues to increase steadily with the number of lifetime hours.

This would be no surprise to Anders Ericsson, the cognitive scientist whose work on expertise—to his annoyance—gave rise to that inaccurate but widely held belief in the magical power of 10,000 hours to bestow mastery.[6] Rather than just the sheer hours of practice put in, Ericsson's research reveals, it's how *smart* those hours are.

What he calls "deliberate" practice involves an expert coach giving feedback on how you are doing, so that you can practice improving in a manner targeted to your progress. A golfer can get pinpointed advice from her coach on exactly how to improve her swing; likewise a surgeon in training, from more seasoned surgeons, on how to improve medical technique. And once the golfer and surgeon have practiced those improvements to the point of mastery, the coaches can give them further feedback for their next round of gains.

This is why so many professional performers—in sports, theater, chess, music, and many other walks of life—continue to have coaches throughout their career. No matter how good you are, you can always get just a bit better. In competitive arenas, small improvements may make all the difference between winning and losing. And if you are not competing, it's your personal best that notches upward.

The same applies to meditation. Take the case of Richie and Dan. We have continued to practice regularly over the decades, for many of those years doing a weeklong retreat or two. We each have sat in meditation every morning for more than forty years (except if something like a 6:00 a.m. flight disrupts the routine). While we both might technically qualify as long-term meditators, with somewhere around 10,000 lifetime hours of practice, neither of us feels particularly evolved when it comes to extremely positive altered traits. Why?

For one, the data suggests that meditating for one session daily is very different from a multiday or longer retreat. Take a finding that emerged unexpectedly in the study of seasoned meditators (9,000 hours average) and their reactivity to stress[7] (see chapter five, "A Mind Undisturbed"). The stronger the connectivity between the meditators' prefrontal area and amygdala, the less reactive they were. The surprise: the greatest increase in prefrontal-amygdala connection correlated with the number of hours a meditator had spent in retreat but not with home hours.

Along these lines another surprising finding was from the study of breath rate. A meditator's hours of retreat practice most strongly correlated with slower breathing, much more than daily practice.[8]

One important difference about meditation on retreat is that there are teachers available who can provide guidance—like a coach. And then there is the sheer intensity of the retreat practice, where meditators typically spend up to eight hours (and sometimes much more) a day in formal practice, often for many days in a row. And many or most retreats are at least partially in silence, which may well contribute to building intensity. All of that adds up to a unique opportunity to amp up the learning curve.

Another difference between amateurs and experts has to do with *how* they practice. Amateurs learn the basic moves of the skill—whether golf, chess, or, presumably, mindfulness and the like—and very often level off after about fifty hours of improving through practice. For the rest of the time their skill level stays about the same—further practice does not lead to great improvements.

Experts, on the other hand, practice differently. They do intensive sessions under the watchful eye of a coach, who suggests to them what to work on next to get even better. This leads to a continuous learning curve with steady improvements.

These findings point to the need for a teacher, someone more advanced than you are, who can give you coaching on how to improve. Both of us have continued to seek guidance from meditation teachers over the years, but the opportunity occurs sporadically in our lives.

The *Visuddhimagga* advises practitioners to find as a guide someone more experienced than they are. This ancient list of potential teachers starts at the top with, ideally, direction from an *arhant* (the Pali word for a fully accomplished meditator, someone at the Olympic level). If none is available, it advises, just find someone more seasoned than you—at the very least, they should have read a *sutra*, a passage from a holy text—if you have read none. In today's world, that might be the equivalent of getting instruction from someone who had tried out a meditation app—it's better than nothing.

BRAIN MATCHING

"Your program," Dan wrote to Jon Kabat-Zinn, "could spread throughout the healthcare system." Little did he know. The year was 1983, and Jon was still working hard just to get doctors at his medical center to send him patients.

Dan was encouraging Jon to do some research on the program's effectiveness—perhaps a small seed of the hundreds of such studies on MBSR today. Dan and Richie, with their thesis adviser at Harvard, had come up with a soft measure of whether people experience anxiety mainly in their mind or in their body. Pointing out that the MBSR program offered both cognitive and somatic practices, Dan suggested Jon study "which elements work best for which type."

Jon went ahead with that study; one finding was that those at the extreme for worries and anxious thoughts (that is, cognitive anxiety) found most relief from doing the yoga in MBSR.[9] This raises a question for all types of meditation—and the more widely deployed user-friendly versions that derive from them: Which forms of practice are most helpful to which people?

Matching the student to the method has ancient roots. In the *Visuddhimagga*, for instance, meditation teachers are advised to carefully observe their students to assess which category they fit in—"greedy" or "hateful" types being two examples—all the better to match them to circumstances and methods most suitable. The matches, which might seem to modern sensibilities a bit medieval, include: for the greedy (who, for instance, first notice what's beautiful), bad food and uncomfortable lodgings and the loathsomeness of body parts as the object of meditation. For the hateful (who first notice what's wrong), the best food and a room with a comfy bed to sleep and meditate on soothing topics like loving-kindness or equanimity.

A more scientifically based optimal matching could start by using existing measures of people's cognitive and emotional styles, as Richie and Cortland Dahl have proposed.[10] For example, for those prone to ruminating and worrying about themselves, a helpful starter practice might be mindfulness of thoughts, where they learn to regard thoughts as "just thought," without getting wrapped up in their content (or yoga, as Jon found). And, perhaps, feedback from their sweat response, a measure of emotional hijacking by thoughts, could further help. Or a person with strong, focused attention but a deficit in empathic concern might begin with compassion practice.

One day those match-ups might be based on a brain scan that helps point people to the optimal method. Such matching of medicine to diagnosis already goes on in some academic medical centers with "precision medicine," where treatments are tailored to an individual's specific genetic makeup.

TYPOLOGIES

Neem Karoli Baba, the remarkable yogi Dan met on his first visit to India, often stayed at Hindu temples and ashrams dedicated to Hanuman, the monkey god. His followers practiced bhakti, the yoga of devotion dominant in the parts of India where he stayed.

While he never talked about his own practice history, bits leaked out now and then. Word had it he had lived for a long time as a jungle yogi; some say he also practiced in an underground cave for years. His meditations were devotional, to Ram, the hero of the Indian epic *Ramayana*; he could sometimes be heard reciting "Ram, Ram, Ram . . ." under his breath, or counting the mantra on his fingers.

He was also said to have traveled to Mecca in the 1930s with a Muslim devotee. To Westerners he praised Christ. For two years he took under his wing, and became close friends with, Lama Norla, who had fled Tibet for India in 1957, long before there were settlements for such refugees. (Lama Norla was a retreat master in one of the very meditation lineages that Mingyur Rinpoche has practiced in.)

If someone was following a given inner path, Neem Karoli always encouraged it. From his perspective the main point was that you do your practice—not try to find the "very best."

Whenever Neem Karoli was asked about which path was best, his answer was *"Sub ek!"*—Hindi for "They are all one." Everyone has different preferences, needs, and the like. Just choose one and plunge in.

In that view contemplative paths are more or less the same, a doorway beyond ordinary experience. At a practical level, all forms of meditation share a common core of mind training—e.g., learning to let go of the myriad distractions that flow through the mind and to focus on one object of attention or stance of awareness.

But as we get more familiar with the mechanics of the various paths, they divide and cluster together. For instance, someone silently reciting a mantra and ignoring everything else deploys different mental operations than does a person who mindfully observes passing thoughts.

And at the finest-grained level, each path in its particulars is quite unique. A student of bhakti yoga who sings devotional *bhajans* to a deity may share some aspects, but not others, with a Vajrayana practitioner who silently generates an image of a deity, like the compassionate Green Tara, along with trying to generate the qualities that go with that image.

We should note that the three levels of practice well studied so far—beginner, long-term, and yogi—are grouped around different kinds of meditation: mainly mindfulness for beginners, vipassana for long-term (with some studies of Zen, too), and for the yogis, the Tibetan paths known as Dzogchen and Mahamudra. As it happens, our own practice history has followed this rough trajectory, and in our experience there are significant differences among these three methods.

Mindfulness, for instance, has the meditator witness whatever thoughts and feelings come and go in the mind. Vipassana starts there, then transitions into a meta-awareness of the processes of mind, not the shifting contents. And Dzogchen and Mahamudra include both in early stages—and a host of other meditation types—but end in a "nondual" stance, resting in a more subtle level of "meta-awareness."

This raises a scientific question about the vector of transformation: Can we extrapolate insights from mindfulness and apply them to vipassana (a traditional segue), and from vipassana to the Tibetan practices?

Taxonomies help science organize such questions, and Dan attempted one for meditation.[11] His immersion in the *Visuddhimagga* offered him a lens for categorizing the bewildering mélange of meditation states and methods he encountered in his wanderings through India. He built a classification around the difference between one-pointed concentration and the more free-floating awareness of mindfulness, a major divide within vipassana practice (and also in the Tibetan paths, but with very different meanings—it gets complicated).

A more inclusive—and more current—typology by Richie with his colleagues Cortland Dahl and Antoine Lutz organizes thinking about meditation "clusters" on the basis of a body of findings in cognitive science and clinical psychology.[12] They see three categories:

- *Attentional.* These meditations focus on training aspects of attention, whether in concentration, as in zeroing in on the breath, a mindful observation of experience, a mantra, or meta-awareness, as in open presence.
- *Constructive.* Cultivating virtuous qualities, like loving-kindness, typifies these methods.
- *Deconstructive.* As with insight practice, these methods use self-observation to pierce the nature of experience. They include "nondual" approaches that shift into a mode where ordinary cognition no longer dominates.

Such a widely inclusive typology makes glaringly clear how meditation research has focused on a narrow subset of methods and ignored the much larger universe of techniques. The bulk of research has been done on MBSR and related mindfulness-based approaches, and there have been many studies of loving-kindness and TM, plus a handful on Zen.

But the many varieties of meditation beyond these may well target their own range of brain circuitry and cultivate their unique set of particular qualities. We hope that as contemplative science grows, researchers will study a wider variety of meditations, not just a small branch of the entire tree. While findings so far are encouraging, there could well be others that we have not yet even a hint of.

The wider the net, the more we will understand about how meditation training shapes the brain and mind. What, for example, might be the benefits of the meditative whirling practice in some schools of Sufism, or the devotional singing in Hinduism's bhakti branch? Or of the analytic meditation practiced by some Tibetan Buddhists, as well as by some schools of Hindu yogis?

But whatever the particulars of a meditation path, they share one goal in common: altered traits.

CHECKLISTS FOR ALTERED TRAITS

About forty reporters, photographers, and TV camera operators were packed into a small basement room, part of the crypt below the main floor of London's Westminster Cathedral. They were there for a press conference with the Dalai Lama, who was about to receive the Templeton Prize—more than a million dollars given each year to recognize an "exceptional contribution to affirming life's spiritual dimension."

Richie and Dan were in London at that press conference to give reporters a backgrounder on the Dalai Lama's lifelong pursuit of scientific knowledge, and his insight that both science and religion share common goals: pursuit of the truth and serving humanity.

In response to the last question at the press conference, the Dalai Lama announced what he would do with the award: immediately give it away. He explained he needs no money—he's a simple monk, and besides, he's a guest of the Indian government, which takes care of all his needs.

So the moment he gets the award he promptly gives one million–plus dollars to Save the Children, in appreciation of their global work with the world's poorest children, and for having helped Tibetan refugees when they fled China. Then he gives what remains to the Mind and Life Institute, and to Emory University for its Tibetan-language program to educate Tibetan monks in science.

We've seen him do the same over and over. His generosity seems spontaneous and without the least regret or holding on to even a tiny bit for himself. Generosity like this, instant and without attachment, marks one of several qualities found in traditional lists of *paramitas* ("completeness or perfection"; literally, "gone to the other shore"), virtuous traits that mark progress in contemplative traditions.

A definitive work on the paramitas, called *The Way of the Bodhisattva*, was written by Shantideva, an eighth-century monk at Nalanda University in India, one of the world's first places of higher learning. The Dalai Lama frequently teaches this text, always acknowledging his debt to his own tutor on it—Khunu Lama, the same humble monk Dan met in Bodh Gaya.

Among the paramitas, embraced by the practice traditions of the yogis who came to Richie's lab, are *generosity*, whether material, like the Dalai Lama giving away his prize money, or even simple presence, giving of oneself; and *ethical conduct,* not harming oneself or others and following guidelines for self-discipline.

Another: *patience,* tolerance, and composure. This also implies a serene equanimity. "Real peace," the Dalai Lama told an MIT audience, "is when your mind goes twenty-four hours a day with no fear, no anxiety."

There's energetic *effort* and diligence; *concentration* and nondistraction; and *wisdom,* in the sense of insights that come via deep meditation practice.

This notion of actualizing the very best in us as lasting traits resonates broadly across spiritual traditions. As we saw in chapter three, "The After Is the Before for the Next During," Greco-Roman philosophers heralded an overlapping set of virtues. And a Sufi saying has it, "Goodness of character is prosperity enough."[13]

Consider the tale of Rabbi Leib, a student of Rabbi Dov Baer, an eighteenth-century Hasidic teacher. In those times students in that tradition mainly studied religious tomes and heard lectures on passages from the Torah, their holy book. But Leib had a different goal.

He had not gone to Dov Baer, his religious mentor, to study texts or hear sermons, Leib said. Rather, he went to "see how he ties his shoes."[14]

In other words, what he sought was to witness and absorb the qualities of being his teacher embodied.

There are intriguing dovetails between the scientific data and the ancient maps to altered traits. For example, an eighteenth-century Tibetan text advises that among the signs of spiritual progress are lovingkindness and strong compassion toward everyone, contentment, and "weak desires."[15]

These qualities seem to match with indicators of brain changes that we have tracked in earlier chapters: amped-up circuitry for empathic concern and parental love, a more relaxed amygdala, and decreased volume of brain circuits associated with attachment.

The yogis who came to Richie's lab all had practiced in a Tibetan tradition that proffers a view that can sometimes be confusing: that we all have Buddha nature, but we simply fail to recognize it. In this view, the nub of practice becomes recognizing intrinsic qualities, what's already present rather than the development of any new inner skill. From this perspective, the remarkable neural and biological findings among the yogis are signs not so much of skill development but rather of this quality of recognition.

Are altered traits add-ons to our nature, or uncovered aspects that were there all along? At this stage in the development of contemplative science it is difficult to weigh in on either side of this debate. There is, however, an increasingly robust corpus of scientific findings showing, for example, that if an infant watches puppets who engage in an altruistic, warmhearted encounter, or ones who are selfish and aggressive, when given the choice of a puppet to reach for, almost all infants choose one of the friendly ones.[16] This natural tendency continues through the toddler years.

These findings are consistent with the view of preexisting virtues like an intrinsic basic goodness, and invite the possibility that training in loving-kindness and compassion involve recognizing early on a core quality that is present and strengthening it. In this sense, practitioners may not be developing a new skill but rather nurturing a basic competence, in much the same way that language is developed.

Whether the whole range of qualities said to be cultivated by different meditation practices is best viewed in this way or more as skill development will be decided by future scientific work. We simply entertain the idea that at least some aspects of meditation practice may be less like learning a new skill, and more akin to recognizing a basic propensity there from the start.

WHAT'S MISSING?

Historically, meditation was not meant to improve our health, relax us, or enhance work success. Although these are the kinds of appeal that has made meditation ubiquitous today, over the centuries such benefits were incidental, unnoted side effects. The true contemplative goal has always been altered traits.

The strongest signs of these qualities are in the group of yogis who came to Richie's lab. This raises a crucial question for understanding how contemplative practice works. Those yogis all practice within a spiritual tradition, in the "deep" mode. Yet most of us in today's world prefer our practice easy (and brief), a pragmatic approach that tends to borrow what works and leave behind the rest.

And quite a lot has been left behind as the world's rich contemplative traditions morphed into user-friendly forms. As meditation migrates from its original setting into popular adaptations, what has been abandoned is ignored or forgotten.

Some important components of contemplative practice are not meditation per se. In the deep paths, meditation represents just one part of a range of means helping to increase self-awareness, gain insights into the subtleties of consciousness, and, ultimately, to achieve a lasting transformation of being. These daunting goals require lifelong dedication.

The yogis who came to Richie's lab all practiced in a Tibetan tradition that holds the ideal that, eventually, people everywhere can be freed from suffering of all sorts—and that the meditator sets out toward this enormous task through mind training. Part of this yogic mind-set involves developing more equanimity toward our own emotional world, as well as the conviction that meditation and related practices can produce lasting transformation—altered traits.

While some of those who follow the "deep" path in the West may themselves hold such convictions, others who train in those same methods do so on a path to renewal—a kind of inner vacation—rather than a lifelong calling. (That said, motivations can change with progress, so that what brought someone to meditation may not be the same goal that keeps them going.)

The sense of a life mission centered on practice numbers among those elements so often left on a far shore, but that may matter greatly. Among others that might, in fact, be crucial for cultivating altered traits at the level found in the yogis:

- An ethical stance, a set of moral guidelines that facilitate the inner changes on the path. Many traditions urge such an inner compass, lest any abilities developed be used for personal gain.
- Altruistic intention, where the practitioner invokes the strong motivation to practice for the benefit of all others, not just oneself.
- Grounded faith, the mind-set that a particular path has value and will lead you to the transformation you seek. Some texts warn against blind faith and urge students to do what we call today "due diligence" in finding a teacher.

- Personalized guidance. A knowledgeable teacher who coaches you on the path, giving you the advice you need to go to the next step. Cognitive science knows that attaining top-level mastery requires such feedback.

- Devotion, a deep appreciation for all the people, principles, and such that make practice possible. Devotion can also be to the qualities of a divine figure, a teacher, or the teacher's altered traits or quality of mind.

- Community. A supportive circle of friends on the path who are themselves dedicated to practice. Contrast that with the isolation of many modern meditators.

- A supportive culture. Traditional Asian cultures have long recognized the value of people who devote their life to transforming themselves to embody virtues of attention, patience, compassion, and so on. Those who work and have families willingly support those who dedicate themselves to deep practice by giving them money, feeding them, and otherwise making life easier. Not so in modern societies.

- Potential for altered traits. The very idea that these practices can lead to a liberation from our ordinary mind states—not just self-improvement—has always framed these practices, fostering respect or reverence for the path and those on it.

We have no way of knowing how any of those "left-behinds" might actually be active ingredients in the altered traits that scientific research has begun to document in the lab.

AWAKENING

Soon after Siddhartha Gautama, the prince-turned-renunciate, had completed his inner journey at Bodh Gaya, he encountered some wandering yogis. Recognizing that Gautama had undergone some kind of remarkable transformation, they asked him, "Are you a god?"

To which he replied, "No. I am awake."

The Sanskrit word for "awake," *bodhi*, gave Gautama the name we know him by today, Buddha—the Awakened One. No one can know with absolute certainty what that awakening entailed, but our data on the most advanced yogis may yield some clues. For instance, there's that high level of ongoing gamma, which seems to lend a sense of vast spaciousness, senses wide-open, enriching everyday experience—even deep sleep, suggesting an around-the-clock quality of awakening.[17]

The metaphor of our ordinary consciousness as a kind of sleep, and an inner shift leading to becoming "awake" has a long history and wide circulation. While various schools of thought contend on the point, we are not prepared nor qualified to wade into the countless debates about what "awakening" means exactly, nor do we contend that science can referee metaphysical debates.

Just as math and poetry are different ways of knowing reality, science and religion represent disparate *magisteria*, realms of authority, areas of inquiry and ways of knowing—religion speaking to values, beliefs, and transcendence, and science to fact, hypotheses, and rationality.[18] In taking the measure of the meditator's mind we do not speak to the truth-value of what various religions make of those mental states.

We aim for something more pragmatic: What in these processes of transformation from the deep path might be extracted that has wide universal benefit? Can we draw on the mechanics of the deep path to create benefits for the widest numbers?

IN A NUTSHELL

From the beginning hours, days, and weeks of meditation, several benefits emerge. For one, beginners' brains show less amygdala reactivity to stress. Improvements in attention after just two weeks of practice include better focus, less mind-wandering, and improved working memory—with a concrete payoff in boosted scores on a graduate school entrance exam. Some of the earliest benefits are with compassion meditation, including increased connectivity in the circuitry for empathy. And markers for inflammation lessen a bit with just thirty hours of practice. While these benefits emerge even with remarkably modest hours of practice, they are likely fragile, and need daily sessions to be sustained.

For long-term meditators, those who have done about 1,000 hours or more of practice, the benefits documented so far are more robust, with some new ones added to the mix. There are brain and hormonal indicators of lowered reactivity to stress and lessened inflammation, a strengthening of the prefrontal circuits for managing distress, and lower levels of the stress hormone cortisol, signaling less reactivity to stresses in general. Compassion meditation at this level brings a greater neural attunement with those who are suffering, and enhanced likelihood of doing something to help.

When it comes to attention, there are a range of benefits: stronger selective attention, decreased attentional blink, greater ease in sustaining attention, a heightened readiness to respond to whatever may come, and less mind-wandering. Along with fewer self-obsessed thoughts comes a weakening of the circuitry for attachment. Other biological and brain changes include a slower breath rate (indicting a slowing of the metabolic rate). A daylong retreat enhances the immune system, and signs of meditative states continue during sleep. All these changes suggest the emergence of altered traits.

Finally, there are the yogis at the "Olympic" level, who have an average of 27,000 lifetime hours of meditation. They show clear signs of altered traits, such as large gamma waves in synchrony among far-flung brain regions—a brain pattern not seen before in anyone—and which also occurs at rest among those yogis who have done the most hours of practice. While strongest during the practices of open presence and of compassion, the gamma continues while the mind is at rest, though to a lesser degree. Also, yogis' brains seem to age more slowly compared to brains of other people their age.

Other signs of the yogis' expertise include stopping and starting meditative states in seconds, and effortlessness in meditation (particularly among the most seasoned). Their pain reaction, too, sets the yogis apart: little sign of anticipatory anxiety, a short but intense reaction during the pain itself, and then a rapid recovery. During compassion meditation, yogis' brains and hearts couple in ways also not seen in other people. Most significant, the yogis' brain states at rest resemble the brain states of others while they meditate—the state has become a trait.

14

A Healthy Mind

D r. Susan Davidson, Richie's wife, is a specialist in high-risk obstetrics—and, like Richie, a longtime meditator. Some years back Susan and a few others decided to organize a meditation group for the doctors in her hospital in Madison. The group met Fridays, in the morning. Susan would send out regular emails to the hospital's physicians reminding them of the opportunity. And very often she would be stopped in the hallway by one or another of them who said, in effect, "I'm so glad you're doing this."

And then add, "But I can't come."

To be sure, there were good reasons. The physicians at the time were even more busy than usual, trying to implement electronic record keeping before there were ready-made templates for it. And the medical specialty that trains "hospitalists"—in-house physicians and staff who deliver comprehensive care to inpatients, freeing up time for others from having to make rounds—did not yet exist. So the meditation group likely would have represented a boon for those harried physicians, a chance to restore themselves a bit.

But still, over the years only six or seven physicians showed up at any given session. Eventually Susan and the others ran out of steam; feeling the group never got real traction, they ended it.

That feeling of not having time may be the number one excuse among people who want to meditate but never get around to it.

Recognizing this, Richie and his team are developing a digital platform called Healthy Minds that teaches meditation-based strategies to cultivate well-being, even for those who say they "have no time." If you insist you are too busy for formal meditation, Healthy Minds can be tailored so you can piggyback your practice on something you do anyway, like commuting or cleaning the house. As long as that activity does not demand your full attention, you can listen to practice instructions in the background. Since some of the main payoffs from meditation are in how they prepare us for everyday life anyway, the chance to practice in the midst of life could be a strength.

Healthy Minds, of course, adds to the ever longer list of apps that teach meditation. But while those many apps use the scientific findings on the benefits of meditation as a selling point, Healthy Minds will go one crucial step further: Richie's lab will scientifically investigate its impacts to assess how well such piggybacked practice actually works.

For example, how does twenty minutes a day during commuting compare to twenty minutes a day sitting in a quiet place at home? We don't know the answer to this simple question. And is it better to practice in a single twenty-minute period, two ten-minute periods, or four periods of five minutes each? These are among the many practical questions Richie and his team hope to answer.

We see this digital platform and the research evaluating it as a prototype of the next step in widening the path of access to the many benefits science finds from contemplative practice. Already MBSR, TM, and generic forms of mindfulness are in easy-to-access forms anyone might benefit from, without having to embrace, or even know about, their Asian roots.

Many companies, for instance, have deployed these approaches as beneficial both for their employees and for the bottom line, offering contemplative methods as part of their training and development menus; some even have meditation rooms where employees can spend quiet time focusing. (Of course, such offerings need a supportive work culture—at one company where workers pounded away at their terminals for exhausting hours on end, Dan was told in confidence that people seen using the meditation room there too often might be fired.)

Amishi Jha's group at the University of Miami now offers mindfulness training to high-stress groups ranging from combat troops to football players, firefighters, and teachers. The Garrison Institute outside New York City offers a mindfulness-based program to help frontline trauma workers in Africa and the Middle East deal with their secondary trauma from, e.g., fighting the Ebola epidemic or helping desperate refugees. And Fleet Maull, while serving a fourteen-year sentence for drug smuggling, founded the Prison Mindfulness Institute, which now teaches inmates in close to eighty prisons across America.

We see contemplative science as a body of basic information about the many ways our minds, bodies, and brains can be molded toward health in its broadest sense. "Health," as the World Health Organization defines it, goes beyond the absence of disease or disability to include "complete physical, mental, and social well-being." Meditation and its derivatives can be an active ingredient in such well-being in several ways, and can have a long reach to far corners.

Findings from contemplative science can spawn innovative approaches that are soundly evidence-based, but which look nothing like meditation per se. These applications of meditation to help solve personal and social dilemmas are all to the good. But what the future might bring excites us, too.

Distancing these methods from their roots may be to the good—so long as what emerges stays grounded in science—making these solutions more readily available to the widest range of people who might benefit. Why, after all, should these methods and their benefits just be for meditators?

GUIDING NEUROPLASTICITY

"What do plants need to grow?" asked Laura Pinger, a curriculum specialist in Richie's center who developed the Kindness Curriculum for preschool children.

That morning, many among the fifteen preschoolers learning to emphasize kindness eagerly waved their hands to answer.

"Sunlight," said one.

"Water," said another.

And a third, who had struggled with attention problems but benefited greatly from the kindness program, shot up his hand and blurted out, "Love."

There was a palpable moment of appreciation for what became a teachable moment. The lesson this led to was about kindness as a form of love.

The Kindness Curriculum begins with very basic, age-appropriate mindfulness exercises where the four-year-olds listen to the sound of a bell and pay attention to their breathing as they lie on their backs, small stones placed on their tummies rising and falling with each breath.

They then use that mindful attention to focus awareness on their body, learning how to pay close attention to those feelings while interacting with other kids—particularly if that other child has gotten upset. Such upsets become opportunities to have the children not only notice what is happening in their own bodies but also imagine what might be happening in the body of their upset classmate—a venture into empathy.

The children are encouraged to practice helping one another, and to express gratitude. When children appreciate the helpfulness of another, they can reward that act by telling the teacher, who will give the helpful child a sticker on a poster of a "kindness garden."

To evaluate the impact of this program, the Davidson group invited children to share stickers (important currency for a toddler) with one of four children: their favorite person in the class; their least favorite classmate; a stranger—a kid they've never met—or a sick-looking child.

Toddlers in the kindness curriculum shared more with the least favorite and the sick children, compared to other kids in standard pre-K who gave most stickers to their favorite person.[1] Another finding: unlike most children, the kindness kids did not become self-focused when they reached kindergarten.

Helping children develop kindness seems an obvious, good idea—but at present this valuable human capacity is left to chance in our educational system. Many families, of course, instill these values in their children—but many do not. Getting such programs into schools ensures that all children will have the lessons that will strengthen this muscle of the heart.[2]

Kindness, caring, and compassion all follow a line of development that our educational system largely ignores—along with attention, self-regulation, empathy, and a capacity for human connection. While we do a good enough job with the traditional academic skills like reading and math, why not expand what children learn to include such crucial skills for living a fulfilled life?

Developmental psychologists tell us that there are differing rates of maturation for attention, for empathy and kindness, for calmness and for social connection. The behavioral signs of this maturation— like the rambunctiousness of kindergartners versus better-behaved fourth graders—are outer signs of growth in underlying neural networks. And neuroplasticity tells us all such brain circuitry can be guided in the best direction through training like the Kindness Curriculum.

At present how our children develop these vital capacities has been left mainly to random forces. We can be smarter in how we help children cultivate them. For instance, all meditation methods at their root are practices in strengthening attention. Adapting these techniques in ways that bring attention-building exercise to children has an array of advantages. No attention, nothing learned.

It's remarkable how little consideration goes to strengthening attention in children, especially because childhood offers a long period of opportunity for growth in the brain's circuitry, and added help might strengthen those circuits. The science of cultivating attention is quite robust, so the path to accomplishing this aim is within our reach.

And we have all the more reason: our society suffers from an attention deficit. Today's children grow up with a digital device at hand continuously, and those devices offer constant distractions (and a larger stream of information than for any generation in the past), so we consider boosting attention skills to be nothing short of an urgent public health need.

Dan was a cofounder of the movement called "social/emotional learning," or SEL; today there are thousands of schools offering SEL around the world. Boosting attention and empathic concern, he has argued, are the next step.[3] To be sure, a robust movement has emerged to bring mindfulness to schools and particularly to poor or troubled youth.[4] But these are isolated efforts or pilots. We envision programs in focusing attention and kindness one day being part of the standard offerings for all children.

Given how much time school-age kids spend playing video games, that speaks to another route to delivering these lessons. The games, to be sure, are sometimes demonized as contributing to the attention deficit we collectively face in modern culture. But imagine a world in which their power can be harnessed for good, for cultivating wholesome states and traits. Richie's group has collaborated with video game designers who specialize in educational games to create some for young teens.[5]

Tenacity is the name of a video game based on research in Richie's lab on breath counting.[6] It turns out that if you are asked to tap an iPad on each in-breath, most people can do this very accurately. However, if they are also asked to tap with two fingers every ninth breath, on this second task they make mistakes, indicating their mind has wandered.

Richie and his colleagues used this information as the core game mechanic in developing Tenacity. Kids tap the iPad with one finger on each in-breath and with two fingers every fifth breath. Since most kids are highly accurate in tapping with each in-breath, Richie's team can determine if the double-finger taps correctly tracked each fifth breath. The more strings of five accurate counts, the higher the score on the game. And with every correct two-fingered tap, the iPad screen's scenery gets more decoration; in one version, gorgeous flowers began to sprout in a desert landscape.

Playing the game for just twenty to thirty minutes daily over two weeks, Richie's group found, increased connectivity between the brain's executive center in the prefrontal cortex and circuitry for focused attention.[7] And in other tests, the players were better able to focus on someone's facial expression and ignore distractions—signs of increased empathy.

No one believes these changes will last without continued practice of some kind (ideally, without the game). But the fact that beneficial changes occurred both in the brain and in behavior is a proof of concept that video games can improve mindful attention and empathy.

THE MENTAL GYM

When Richie gave that high-profile lecture at the National Institutes of Health, the in-house notice of his talk offered this intriguing speculation: "What if we could exercise our minds like we exercise our bodies?"

The fitness industry thrives on our wish to be healthy; physical fitness is a goal most everyone espouses (whether or not we do much about it). And habits of personal hygiene like regular bathing and tooth brushing are second nature. So why not mental fitness?

Neuroplasticity—the shaping of the brain by repeated experiences—goes on unwittingly throughout our days, though we are typically unaware of these forces. We spend long hours ingesting what's on the screen of our digital devices, or in countless other relatively mindless pursuits. Meanwhile our neurons are dutifully strengthening or weakening the relevant brain circuitry. Such a haphazard mental diet most likely leads to equally haphazard changes in the muscle of the mind.

Contemplative science tells us we can take more responsibility for the care of our own minds. The benefits from shaping our minds more intentionally can come early, as we saw in the data on loving-kindness practices.

Consider work by Tracy Shors, a neuroscientist who developed a training program she hypothesized would increase neurogenesis—the growth of new brain cells—called Mental and Physical (MAP) Training.[8] Participants did thirty minutes of focused attention meditation followed by thirty minutes of moderate-intensity aerobic exercise two times a week for eight weeks. Benefits included improved executive function, supporting the notion that the brain was shaped positively.

While working out intensively produces more muscle and better endurance, if we stop exercising, we know that we are heading back toward more breathlessness and flab. The same goes for the changes in the mind and brain from that inner workout, meditation and its spinoffs.

And since the brain is like a muscle that improves with exercise, why not an equivalent of physical fitness programs—mental gyms? The mental gym would not be a physical space but rather a set of apps for inner exercises that can be performed anywhere.

Digital delivery systems can offer the benefits of contemplative practice to the very widest numbers. While meditation apps are already in wide use, there are no direct scientific evaluations of these methods. Instead the apps typically cite studies done elsewhere on some kind of meditation (and not necessarily the best such studies), while failing to be transparent about their own effectiveness. One such app, which supposedly enhanced mental functions, had to pay a large fine when government agencies challenged their claims, which proved unsupported.

On the other hand, the evidence so far suggests that were well-designed digital deliveries to be tested with rigor, they might do well. For example, there was that study of web-based instruction in loving-kindness (reviewed in chapter six, "Primed for Love") that showed it made people both more relaxed and more generous.[9]

And Sona Dimidjian's group reached out on the web to people who have low-level symptoms of depression—a group at a higher-than-average risk for a bout of full-blown depression. Sona's team developed a web-based course, derived from MBCT, called Mindful Mood Balance; the eight sessions reduced symptoms of depression and anxiety such as constant worry and rumination.[10]

But these success stories do not automatically mean any and all online teaching of meditation or its derivatives will be beneficial. Are some more effective than others? If so, why? These are empirical questions.

To the best of our knowledge, there is not a single publication in the mainstream scientific literature that has directly evaluated the efficacy of any of the multitude of meditation apps that claim a basis in science. We hope one day such an evaluation would be standard for any such app, to show it works as promised.

Still, meditation research offers abundant support for the likely payoff from mind training. We envision a time when our culture treats the mind in the same way it treats the body, with exercises to care for our mind becoming part of our daily routine.

NEURAL HACKING

The New England snow was somewhere between icy and melting that March morning, and the living room of the Victorian house on the Amherst College campus contained a small Noah's ark of disciplines. There were pairs of religious scholars, experimental psychologists, neuroscientists, and philosophers.

The group had gathered under the auspices of the Mind and Life Institute to explore the corner of the mind that begins with everyday desire. Sometimes that pathway runs through craving to addiction— be it to drugs, porn, or shopping.

The religious scholars there pinpointed the problem at the moment of grasping, the emotional impulse that makes us lean in toward pleasure, whatever form it might take. In the grip of grasping, particularly as it slides in intensity toward craving and addiction, there's a sense of uneasiness that drives the clinging, seductive mental whispers that the particular object of our desires will relieve our dis-ease.

Moments of grasping can be so subtle they pass by unnoticed in the frenetic distractions of our usual state of mind. We are most likely, the research shows, to reach for that fattening treat in the moments we are most distracted—and addicts are likely to seek the next fix when they see small prompts, like the shirt they wore during a high, that flood them with memories of their last fix.

This state contrasts, philosopher Jake Davis noted, with the sense of utter ease we feel when we are free of compulsive motivations.

A "mind of nongrasping" renders us immune to these impulses, content in ourselves as we are.

Mindfulness lets us observe what's happening in the mind itself rather than simply be carried away by it. Those impulses to grab start to stand out. "You need to see it to let it go," said Davis. While we are mindful we notice such impulses arising but regard them in the same way as other spontaneously arising thoughts.

The neural action here revolves around the PCC (postcingulate cortex), suggested psychiatrist and neuroscientist Judson Brewer, who had just become director of research at the Center for Mindfulness at the University of Massachusetts Medical School in Worcester— birthplace of MBSR. Mental activities where the PCC plays a part include being distracted, letting our mind wander, thinking about ourselves, liking a choice we've made even if we find it immoral, and feeling guilty. And, oh yes—craving.

Brewer's group, as we saw in chapter eight, "Lightness of Being," imaged the brains of people during mindfulness, finding the method quiets the PCC. The more effortless mindfulness becomes, the quieter the PCC.[11] In Brewer's lab, mindfulness has helped people addicted to cigarettes kick the habit.[12] He has developed two apps—for overeating and for smoking—applying his PCC findings to breaking addictions.

Brewer went on to translate this neural finding into a practical approach using "neurofeedback," which monitors the activity of a person's brain and tells them instantly if a given region is getting more or less active. This allows the person to experiment with what their mind can do to make their PCC less active. Ordinarily we are oblivious to what goes on within our brain, particularly at the level read by brain scanners and the like. That's a main reason neuroscience findings carry such weight. But neurofeedback pierces that mind-brain barrier, opening a window on the brain's activity to allow a feedback loop. This lets us sense how a given mental maneuver impacts the goings-on within our brain. We envision a next generation of meditation-derived apps that use feedback from relevant biological or neural processes, with Brewer's PCC neurofeedback as a prototype.

Another target for neurofeedback might be gamma waves, that EEG pattern that typifies the brain of advanced yogis. Still, while some gamma wave feedback simulation of a yogi's vast openness might result, we do not see neurofeedback as a shortcut to the yogi's realization of altered traits. Gamma oscillations, or any particular measure taken of the yogis' state of mind, offers at best an arbitrary and thin slice of the rich fullness yogis seem to enjoy. While gamma wave feedback, or some other dip into such elements, may offer a contrast to our ordinary mind states, they by no means equate with the fruits of years of contemplative practice.

But there are other possible payoffs. Consider the meditating mice.

Meditating Mice?!? This ridiculous possibility—or a very vague parallel—has been explored by neuroscientists at the University of Oregon. Okay, the mice didn't really meditate; researchers used a specialized strobe light to drive the mouse's brain at specific frequencies, a method called photic driving, where the rhythm of EEG waves lock into that of a flashing bright light. The mice seemed to find this relaxing, judging from rodent signs of lessened anxiety.[13] When other researchers drove the rodent brain into the gamma frequency with photic driving, they found it reduced the neural plaque associated with Alzheimer's disease, at least in aged mice.[14]

Could feedback of gamma waves (that frequency abundant in yogis) slow or reverse Alzheimer's disease? The annals of pharmaceutical research are rife with potential medications that seemed successful when used in mice, but failed once human trials began.[15] Gamma wave neurofeedback for preventing Alzheimer's disease in humans may (or may not) be a pipe dream.

But the basic model, that neurofeedback apps may make once rarefied states available to a wide swath of people, seems more promising. Here again we see caveats—not the least being that such devices are likely to produce temporary state effects, not lasting traits. Let alone the huge divide between years of intensive meditation and merely using a new app for a bit.

Still, we envision a next generation of helpful applications, all derived from the methods and insights unveiled by contemplative science. What shapes these will eventually take we just do not know.

OUR JOURNEY

The hard evidence for altered traits came slowly, over decades. We were graduate students when we started on the scent, and now, as we sum up what has, finally, become compelling evidence, have reached the era of life when people look toward retiring.

For much of this time we had to pursue a scientific hunch with few supporting data. But we were comforted by the dictum that "an absence of evidence is not evidence of absence." The roots of our conviction lay in our own experiences in meditation retreats, the few rare beings we had met who seemed to embody altered traits, and our reading of meditation texts that pointed to these positive transformations of being.

Still, from an academic point of view, this amounted to an absence of evidence: there were no impartial empirical data. When we began this scientific journey there were scant methods available to explore altered traits. In the 1970s, we were stymied—we could only do studies that tangentially spoke to the idea. For one thing, we had no access to the appropriate subjects—instead of dedicated yogis from remote mountain hermitages, we had to settle for Harvard sophomores.

Most important, human neuroscience was in its tentative, beginning phase. The methods at hand for studying the brain were primitive by today's standards; "state of the art" in those days meant vague or indirect measures of brain activity.

In the decade before our Harvard years philosopher Thomas Kuhn published *The Structure of Scientific Revolutions*, holding that science shifts abruptly from time to time as novel ideas and radically innovative paradigms force shifts in thinking. This idea had caught our fancy as we searched for paradigms that posited human possibilities undreamt of in our psychology. Kuhn's ideas, hotly discussed in the scientific world, spurred us on despite opposition from our own faculty advisers.

Science needs its adventurers. That's what we were as Richie sat on his zafu through that hour of not moving with Goenka-ji, and what Dan was as he hung out with yogis and lamas, and spent months poring over that fifth-century guidebook for meditators, the *Visuddhimagga*.

Our conviction regarding altered traits made us vigilant for studies that might support our hunch. We filtered the findings through the lens of our experience, drawing out implications few others, if any, were seeing.

Sciences operate within a web of culture-bound assumptions that limit our view of what is possible, most powerfully for the behavioral sciences. Modern psychology had not known that Eastern systems offer means to transform a person's very being. When we looked through that alternate Eastern lens, we saw fresh possibilities.

By now, mounting empirical studies confirm our early hunches: sustained mind training alters the brain both structurally and functionally, proof of concept for the neural basis of altered traits that practitioners' texts have described for millennia. What's more, we all can move along this spectrum, which seems to follow a rough dose-response algorithm, gaining benefits in accord with our efforts.

Contemplative neuroscience, the emerging specialty which supplies the science behind altered traits, has reached maturity.

CODA

"What if, by transforming our minds, we could improve not only our own health and well-being but also those of our communities and the wider world?"

That rhetorical question, too, comes from the internal notice at the National Institutes of Health about Richie's talk there.

So, what if?

We envision a world where widespread mental fitness deeply alters society for the better. We hope the scientific case we make here shows the enormous potential for enduring well-being from caring for our minds and brains, and convinces you that a little daily mental exercise can go a long way toward the cultivation of that well-being.

Signs of such flourishing include increasing generosity, kindness and focus, and a less rigid division between "us" and "them." In light of increases in empathy and perspective taking from various kinds of meditation, we think it likely that these practices will produce a greater sense of our interdependence on one another and with the planet.

When nurtured on a grand scale, these qualities—particularly kindness and compassion—would inevitably lead to changes for the better in our communities, our nations, and our societies. These positive altered traits have the potential for transforming our world in ways that will enhance not only our individual thriving but also the odds for our species' survival.

We are inspired by the vision of the Dalai Lama as he reached eighty years of age. He encourages us all to do three things: gain composure, adopt a moral rudder of compassion, and act to better the world. The first, inner calm, and the second, navigating with compassion, can be products of meditation practice, as can executing the third, via skillful action. Exactly what action we take, though, remains up to each of us, and depends on our individual abilities and possibilities—we each can be agents in a force for good.[16]

We view this "curriculum" as one solution to an urgent public health need: reducing greed, selfishness, us/them thinking and impending eco-calamities, and promoting more kindness, clarity, and calm. Targeting and upgrading these human capacities directly could help break the cycle of some otherwise intractable social maladies, like ongoing poverty, intergroup hatreds, and mindlessness about our planet's well-being.[17]

To be sure, there are still many, many questions about how altered traits occur, and much more research is needed. But the scientific data supporting altered traits have come together to the point that any reasonable scientist would agree that this inner shift seems possible. Yet too few of us at present realize this, let alone entertain the possibility for ourselves.

The scientific data, while necessary, are by no means sufficient for the change we envision. In a world growing more fractured and endangered, we need an alternative to mind-sets snarky and cynical, views fostered by focusing on the bad that happens each day rather than the far more numerous acts of goodness. In short, we have ever greater need for the human qualities altered traits foster.

We need more people of goodwill, who are more tolerant and patient, more kind and compassionate. And these can become qualities not just espoused but embodied.

We—along with legions of fellow journeyers—have been exploring altered traits, in the field, in the lab, and in our own minds, for more than forty years. So, why this book now?

Simple. We feel that the more these upgrades in the brain, mind, and being are pursued, the more they can change the world for the better. What sets this strategy for human betterment apart from the long history of failed utopian schemes comes down to the science.

We have shown the evidence that it is possible to cultivate these positive qualities in the depths of our being, and that any of us can begin this inner journey. Many of us may not be able to put forth the intense effort needed to walk the deep path. But the wider routes show that qualities like equanimity and compassion are learnable skills, ones we can teach our children and improve in ourselves.

Any steps we take in this direction are a positive offering to our lives and our world.

Further Resources

FOR ONGOING REPORTS OF MEDITATION RESEARCH

https://centerhealthyminds.org/—Center for Healthy Minds, University of Wisconsin–Madison
https://www.mindandlife.org/—Mind & Life Institute
https://nccih.nih.gov/—National Center for Complementary and Integrative Health
http://ccare.stanford.edu/—Center for Compassion and Altruism Research and Education, Stanford University
http://mbct.com/—Mindfulness-based cognitive therapy

KEY MEDITATION RESEARCH GROUPS

https://centerhealthyminds.org/science/studies—Richie Davidson's lab
http://www.umassmed.edu/cfm/—Judson Brewer's lab, and the center for MBSR
https://www.resource-project.org/en/home.html—Tania Singer's meditation study
http://www.amishi.com/lab/—Amishi Jha's lab
http://saronlab.ucdavis.edu/—Clifford Saron's lab
https://www.psych.ox.ac.uk/research/mindfulness—Oxford Mindfulness Centre
http://marc.ucla.edu/—UCLA Mindful Awareness Research Center

SOCIETAL IMPLICATIONS

Dalai Lama's Vision: www.joinaforce4good.org

FOR THE AUDIO VERSION OF THIS BOOK

www.MoreThanSound.net

Acknowledgments

We could not have begun the journey that resulted in this book without the initial inspiration from those spiritually advanced beings we met who have progressed far on the path of meditation.

There are those Dan met in Asia, including Neem Karoli Baba, Khunu Lama, and Ananda Mayee Ma, among several others. And our teachers: S. N. Goenka, Munindra-ji, Sayadaw U Pandita, Nyoshul Khen, Adeu Rinpoche, Tulku Urgyen, and his sons, all rinpoches, too: Chokyi Nyima, Tsikey Chokling, Tsokryi, and, of course, Mingyur.

Then there are the many Tibetan yogis who traveled far to be studied in Richie's lab, as well as the Western retreatants from their center in Dordogne, France. We are deeply indebted to Matthieu Ricard, who bridged the worlds of science and contemplation, making this line of research possible.

The scientists who have contributed their studies to the ever-building mass of contemplative research are too numerous to name, but we are grateful for their scientific work. Special thanks to those at Richie's lab, notably Antoine Lutz, Cortland Dahl, John Dunne, Melissa Rosenkranz, Heleen Slagter, Helen Weng, and many others too numerous to list who together contributed enormously to this work. The work in Richie's center would not be possible without the tireless contributions of the extraordinary administrative staff and leadership, especially Isa Dolski, Susan Jensen, and Barb Mathison.

Among the many friends and colleagues who have made insightful suggestions along the way, we thank Jack Kornfield, Joseph Goldstein, Dawa Tarchin Phillips, Tania Singer, Avideh Shashaani, Sharon Salzberg, Mirabhai Bush, and Larry Brilliant, to mention a few.

And of course we could not have written this book without the loving support and encouragement of our wives, Susan and Tara.

Our biggest debt of gratitude goes to His Holiness the Dalai Lama, who both inspired us by his very being and also pointedly suggested how meditation research could bring the value of these practices to the widest number of people.

Notes

CHAPTER ONE. THE DEEP PATH AND THE WIDE

1. He was probably referring to the expletives that sometimes explode from those with Tourette's syndrome, not obsessive-compulsive disorder, but in the early 1970s clinical psychology was not yet familiar with the Tourette's diagnosis.
2. www.mindandlife.org.
3. Daniel Goleman, *Destructive Emotions: How Can We Overcome Them?* (New York: Bantam, 2003). Also see www.mindandlife.org.
4. The lab was run by our physiology professor, David Shapiro. Among others in the research group were Jon Kabat-Zinn, who was about to begin teaching what has become mindfulness-based stress reduction, and Richard Surwit, then a psychology intern at Massachusetts Mental Health Center, who later became a professor in psychiatry and behavioral medicine at Duke University Medical School. David Shapiro left Harvard to join the faculty at UCLA, where among other topics he studied the physiological benefits of yoga.
5. The key words used in this search were: *meditation, mindfulness meditation, compassion meditation,* and *loving-kindness meditation.*

CHAPTER TWO. ANCIENT CLUES

1. For a kaleidoscopic sense of Neem Karoli Baba as seen through the eyes of Westerners who knew him, see: Parvati Markus, *Love Everyone: The Transcendent Wisdom of Neem Karoli Baba Told Through the Stories of the Westerners Whose Lives He Transformed* (San Francisco: HarperOne, 2015).
2. Mirka Knaster, *Living This Life Fully: Stories and Teachings of Munindra* (Boston: Shambhala, 2010).
3. The throng of meditators included others who had been with Maharaji, including Krishna Das, and Ram Dass himself. Others, including Sharon Salzberg,

John Travis, and Wes Nisker, became vipassana teachers themselves. Mirabai Bush, another attendee, later founded the Center for Contemplative Mind in Society, an organization dedicated to encouraging contemplative pedagogy at the college level, and she codesigned the first course on mindfulness and emotional intelligence at Google.

4. To be sure, some parts of such texts seemed too fanciful to merit serious attention—notably, on achieving supernormal powers—which matched fairly well a similar section in Patanjali's *Yoga Sutras*. These texts dismiss such "powers," like hearing at a great distance, as having no spiritual significance—and indeed, in some Indian epics like the Ramayana, the villains are said to have attained such powers through years of ascetic meditation practices, but without a protective ethical framework (and thus their villainy).

5. See Daniel Goleman, "The Buddha on Meditation and States of Consciousness, Part I: The Teachings," *Journal of Transpersonal Psychology* 4:1 (1972): 1–44.

6. Daniel Goleman, "Meditation as Meta-Therapy: Hypotheses Toward a Proposed Fifth Stage of Consciousness," *Journal of Transpersonal Psychology* 3:1 (1971): 1–25. On reading this once again some forty years later, Dan feels both embarrassed in many ways by its naiveté and, in a few respects, pleased at its prescience.

7. B. K. Anand et al., "Some Aspects of EEG Studies in Yogis," *EEG and Clinical Neurophysiology* 13 (1961): 452–56. Besides being an anecdotal report, this study occurred long before the advent of computerized data analysis.

8. The key notion of Skinner's "radical behaviorism" was that all human activity resulted from learned associations of a given stimulus (famously, Pavlov ringing a bell) and a specific response (a dog salivating in response to the bell!) that gets reinforced (initially by food).

9. The chairman of Richie's department had gotten his PhD from Harvard under B. F. Skinner himself, and brought to NYU his studies of pigeon training via conditioning—along with a labful of caged pigeons. The department chairman was not just rigid in his behaviorist outlook but, in Richie's view, far too adamant, if not outright rabid. In those years behaviorism had taken over many prestigious psychology departments as part of a more general movement in academic psychology to make the field more "scientific" through experimental research—a reaction to the psychoanalytic theories that had dominated the field (which were largely supported by clinical anecdotes rather than experimentation).

10. As a student in the chairman's senior honors seminar, Richie was horrified to find that the text was Skinner's 1957 book *Verbal Behavior,* which claimed that all human habits were learned through reinforcement, with language as the case in point. Some years earlier Skinner's book had come under fierce, and highly visible, attack in a critical review by MIT linguist Noam Chomsky. The critique points out, for instance, that no matter how much a dog heard human language, no amount of reward would get it to talk—while human babies everywhere learn

to do so with no particular reinforcement. This suggests that inherent cognitive abilities, not mere learned associations, propel the mastery of language. For his seminar presentation Richie recapped Noam Chomsky's critique of Skinner's book—and thereafter felt his department chairman worked ceaselessly to undermine him and in fact wanted to throw him out of the department. That seminar drove Richie nuts; he had fantasies about going into the chairman's lab at three in the morning and liberating the pigeons. See Noam Chomsky, "The Case Against Behaviorism," *New York Review of Books*, December 30, 1971.

11. Richie's adviser Judith Rodin had herself just finished her PhD at Columbia University. Rodin went on to a distinguished career in psychology, becoming the dean of the Graduate School of Arts and Science at Yale, later the university's provost, and then the first female president of an Ivy League college, the University of Pennsylvania. As of this writing she is just stepping down as president of the Rockefeller Foundation.

12. For just such methods he turned to John Antrobus, who taught across town at the City College of New York. Richie would hang out in Antrobus's lab, a refuge from the atmosphere of his own department.

13. Daniel Goleman, *Emotional Intelligence* (New York: Bantam, 1995).

14. William James, *The Varieties of Religious Experience* (CreateSpace Independent Publishing Platform, 2013), p. 388.

15. Freud and Rolland: see Sigmund Freud, *Civilization and Its Discontents*. Later, though, transcendental experiences were included in the theories of Abraham Maslow, who called them "peak experiences." From the 1970s there was a nascent movement at the edge of the already peripheral humanistic psychology movement, called "transpersonal" psychology, which took altered states seriously (Dan was an early president of the Association for Transpersonal Psychology). Dan published his first meditation articles in *The Journal of Transpersonal Psychology*.

16. Charles Tart, ed., *Altered States of Consciousness* (New York: Harper & Row, 1969).

17. The excitement and cultural fascination with psychedelics was in a sense an offshoot of the state of brain science at the time, which for years had been advancing its knowledge of neurotransmitters. Dozens of these had been identified in the early 1970s, though their functions were little understood. Forty years later we can identify more than a hundred, with a vastly more sophisticated list of what they do in the brain, along with a healthy appreciation of the complexity of their interactions.

18. A Social Science Research Council fellowship to study the psychological systems within Asian spiritual traditions.

19. This definition of *mindfulness* comes from Nyanaponika, *The Power of Mindfulness* (Kandy, Sri Lanka: Buddhist Publication Society, 1986).

20. Luria Castell Dickinson, quoted in Sheila Weller, "Suddenly That Summer," *Vanity Fair*, July, 2012, p. 72. Similarly, neurologist Oliver Sacks wrote about his

own explorations with a wide range of mind-altering drugs, "some people can reach transcendent states through meditation or similar trance-inducing techniques. But drugs offer a shortcut; they promise transcendence on demand." Oliver Sacks, "Altered States," *The New Yorker,* August 27, 2012, p. 40. While drugs can induce altered states, they do not help with altered traits.

CHAPTER THREE. THE AFTER IS THE BEFORE FOR THE NEXT DURING

1. Healthy and unhealthy: In the academic vernacular of translations, the two are usually called "unwholesome" and "wholesome" "mental factors."

2. Nyanaponika's original name was Siegmund Feniger. He was born Jewish in Germany in 1901, was already a Buddhist in his twenties, and found the writings of another German-born Buddhist, Nyanatiloka Thera (Anton Gueth), particularly inspiring. With the rise of Hitler, Feniger traveled to what was then Ceylon to join Nyanatiloka at a monastery near Colombo. Nyanatiloka had studied meditation with a Burmese monk reputed to be enlightened (that is, an *arhant*) and Nyanaponika later studied with the legendary Burmese meditation master and scholar Mahasi Sayadaw, who was Munindra's teacher.

3. The course also attracted a number of nonstudents, including Mitch Kapor, who later founded Lotus, an early software success.

4. Another teaching assistant who went on to an illustrious career was Shoshanah Zuboff, who became a professor at Harvard Business School, and wrote *In The Age of the Smart Machine* (Basic Books, 1989), among other books. A student, Joel McCleary, became a member of the Jimmy Carter administration and was crucial in getting State Department approval for the Dalai Lama to visit the United States for the first time.

5. The millions who practice yoga in modern centers are not duplicating the standard methods of Asian yogis who even today seek remote places to practice their methods in privacy. Traditionally, teaching these practices involves a single teacher (or "guru") and student, not a class in a yoga studio. And the sets of poses typical in modern settings differ in key ways from the traditional yogic practices: the standing poses were a recent innovation, the format of sets of poses was borrowed from European exercise routines, and yogis in the wild deploy much more *pranayama* to calm the mind and trigger meditative states than is the case in, say, yoga programs designed for fitness rather than to support long sitting sessions in meditation (which was an original purpose of the yoga *asanas*). See William Broad, *The Science of Yoga* (New York: Simon & Schuster, 2012).

6. Richard J. Davidson and Daniel J. Goleman, "The Role of Attention in Meditation and Hypnosis: A Psychobiological Perspective on Transformations of Consciousness," *International Journal of Clinical and Experimental Hypnosis* 25:4 (1977): 291–308.

7. David Hull, *Science as a Process* (Chicago: University of Chicago Press, 1990).

8. Joseph Schumpeter, *History of Economic Analysis* (New York: Oxford University Press, 1996), p. 41.

9. These were the years when the field of neuroscience was just forming, largely based on research with animals, not people. The Society for Neuroscience held its first meeting in 1971. Richie's first meeting was the society's fifth.

10. E. L. Bennett et al., "Rat Brain: Effects of Environmental Enrichment on Wet and Dry Weights," *Science* 163:3869 (1969): 825–26. http://www.sciencemag.org /content/163/3869/825.short. We now know that the growth might also include adding new neurons.

11. For recent reviews of how music training shapes the brain, see C. Pantev and S. C. Herholz, "Plasticity of the Human Auditory Cortex Related to Musical Training," *Neuroscience Biobehavioral Review* 35:10 (2011): 2140–54; doi:10.1016 /j.neubiorev.2011.06.010; S. C. Herholz and R. J. Zatorre, "Musical Training as a Framework for Brain Plasticity: Behavior, Function, and Structure," *Neuron* 2012: 76(3): 486–502; doi:10.1016/j.neuron.2012.10.011.

12. T. Elbert et al., "Increased Cortical Representation of the Fingers of the Left Hand in String Players," *Science* 270: 5234 (1995): 305–7; doi:10.1126/science.270.5234 .305. Six violists, two cellists, and one guitarist, along with six age-matched non-musician controls formed the subjects for one of the most influential studies on the impact of musical training on the brain. The musicians' training ranged from a low of seven years to a high of seventeen years of training. The nonmusicians were age- and gender-matched to the musicians. Importantly, all the musicians played a string instrument and all were right-handed; the left hand of these musicians is continuously engaged in fingering the instrument when they play. Playing a string instrument requires considerable manual dexterity and cultivates enhanced tactile sensitivity that is key to skillful performance. Using a technique to measure the magnetic signals generated by the brain, very much like measuring the electrical signals (though with greater spatial resolution) showed that the size of the cortical surface devoted to representing the fingers of the left hand was dramatically larger in the musicians compared to the nonmusicians. The size of this area was greatest for those musicians who began their training earlier in life.

13. Technically, this is parafoveal vision. The fovea is the area of the retina that receives input from objects just in front of you, while information that is far off to the right or left is parafoveal.

14. Neville studied ten profoundly and congenitally deaf individuals with an average age of thirty and compared them to an age- and gender-matched typically developing group with no hearing deficits. Neville's team tested them on a task that was designed to assess their parafoveal vision. Yellow flashing circles were presented on the screen, with some flashing quickly but most flashing more slowly. The participants' task was to press a button when they saw the less frequently presented faster-flashing yellow circle. Sometimes the circles appeared toward the center of the screen and at other times the circles were

presented toward the sides, in parafoveal vision. The deaf participants were more accurate than the typically developing controls in detecting the yellow circles when they appeared in the periphery. This finding was to be expected since the deaf subjects were all experienced in sign language and thus their visual experience was quite different than controls and included regular exposure to rich information that was not centrally located. But the finding that was most startling was that the primary auditory cortex, the sector of cortical real estate that receives the initial upstream input that begins in the ear, showed robust activation in response to the circles presented off to the side, but only in the deaf subjects. The hearing subjects showed absolutely no activation of this primary auditory region in response to visual input. See G. D. Scott, C. M. Karns, M. W. Dow, C. Stevens, H. J. Neville, "Enhanced Peripheral Visual Processing in Congenitally Deaf Humans Is Supported by Multiple Brain Regions, Including Primary Auditory Cortex," *Frontiers in Human Neuroscience* 2014:8 (March): 1–9; doi:10.3389/fnhum.2014.00177.

15. This research puts to rest a neuro-myth, that in a nephrology-like map of the brain, each area has a specific set of functions, and these cannot change.

16. The very idea posed a grave challenge to a host of hallowed assumptions in psychology—for instance, that by early adulthood personality becomes fixed, and that the person you are at that point would be who you were for the rest of your life—personality was stable across time and in different contexts. Neuroplasticity suggested otherwise, that your life experience could alter your personality traits to some extent.

17. See, e.g., Dennis Charney et al., "Psychobiologic Mechanisms of Post-Traumatic Stress Disorder," *Archives of General Psychiatry* 50 (1993): 294–305.

18. D. Palitsky et al., "The Association between Adult Attachment Style, Mental Disorders, and Suicidality," *Journal of Nervous and Mental Disease* 201:7 (2013): 579–86; doi:10.1097/NMD.0b013e31829829ab.

19. More formally, an altered trait represents sustained, beneficial qualities of thinking, feeling, and acting resulting from purposeful mind training and accompanied by lasting, supportive changes in the brain.

20. Cortland Dahl et al., "Meditation and the Cultivation of Wellbeing: Historical Roots and Contemporary Science," *Psychological Bulletin*, in press, 2016.

21. Carol Ryff interviewed at http://blogs.plos.org/neuroanthropology/2012/07/19/psychologist-carol-ryff-on-wellbeing-and-aging-the-fpr-interview/.

22. Rosemary Kobau et al., "Well-Being Assessment: An Evaluation of Well-Being Scales for Public Health and Population Estimates of Well-Being among US Adults," *Applied Psychology: Health and Well-Being* 2:3 (2010): 272–97.

23. Viktor Frankl, *Man's Search for Meaning* (Boston: Beacon Press, 2006).

24. Tonya Jacobs et al., "Intensive Meditation Training, Immune Cell Telomerase Activity, and Psychological Mediators," *Psychoneuroendocrinology* 2010; doi: 10.1016/j.psyneurn.2010.09.010.

25. Omar Singleton et al., "Change in Brainstem Gray Matter Concentration Following a Mindfulness-Based Intervention is Correlated with Improvement in Psychological Well-Being," *Frontiers in Human Neuroscience*, February 18, 2014; doi: 10.3389/fnhum.2014.00033.

26. Shauna Shapiro et al., "The Moderation of Mindfulness-Based Stress Reduction Effects by Trait Mindfulness: Results from a Randomized Controlled Trial," *Journal of Clinical Psychology* 67:3 (2011): 267–77.

CHAPTER FOUR. THE BEST WE HAD

1. Richard Lazarus, *Stress, Appraisal and Coping* (New York: Springer, 1984).

2. Daniel Goleman, "Meditation and Stress Reactivity," Harvard University PhD thesis, 1973; Daniel Goleman and Gary E. Schwartz, "Meditation as an Intervention in Stress Reactivity," *Journal of Consulting and Clinical Psychology* 44:3 (June 1976): 456–66; http://dx.doi.org/10.1037/0022-006X.44.3.456.

3. Daniel T. Gilbert et al., "Comment on 'Estimating the Reproducibility of Psychological Science,'" *Science* 351:6277 (2016); doi 10.1126/science.aad7243.

4. The self-assessment Dan used, the State-Trait Anxiety Measure, continues to be widely deployed in research on stress and anxiety, including in meditation studies. Charles. D. Spielberger et al., *Manual for the State-Trait Anxiety Inventory* (Palo Alto, CA: Consulting Psychologists Press, 1983).

5. Urged by his adviser, Dan spent weeks and weeks studying tomes in the Baker Library at Harvard Medical School to track the brain wiring that leads to the GSR, a burst of sweat on the skin—which at the time was a circuit not yet pieced together from the bits and pieces known about neuroanatomy. Dan's adviser had dreams of publishing a journal article on this—though it never came to pass.

6. To be sure, Richie's main electrical measures were advanced for the time. But even reading the recordings then current gave an imprecise sense of what's actually going on inside the brain, especially compared to contemporary systems for analyzing EEG.

7. Worse, in Dan's study even those peripheral measures were botched to some degree. Besides heart rate and sweat response, Dan had measured EMG, or electromyogram, assessing the level of tension in the frontalis muscle (which knits our brows together when we frown or worry). But the EMG results had to be thrown out because Dan was given erroneous advice about the kind of paste to use to attach these sensors to the forehead.

8. Dan's adviser instructed him to skip the heart rate measure for his dissertation. Only later, for their coauthored paper in an academic journal, did his adviser wangle some funds from the department that allowed hiring some undergrads to do the scoring. But there weren't enough funds to score heart rate for the entire time of recording—only for certain periods Dan's adviser chose as critical—e.g., the slope of recovery from the shop accidents. But here again there was a problem:

the meditators had a stronger reaction to the accidents than did the controls. Though their slope of recovery was steeper—indicating a more rapid return to baseline—this measure did not show them becoming even more relaxed post-accident than the controls. This was a weak point, as noted in later critiques of the study. See, e.g., David S. Holmes, "Meditation and Somatic Arousal Reduction: A Review of the Experimental Evidence," *American Psychologist* 39:1 (1984): 1–10.

9. The crucial comparison pointing to a possible trait effect would be between the seasoned meditators and the novices in the condition where neither group meditated before seeing the accident film.

10. Joseph Henrich et al., "Most People Are Not WEIRD," *Nature* 466:28 (2010). Published online June 30, 2010; doi:10.1038/466029a.

11. Anna-Lena Lumma et al., "Is Meditation Always Relaxing? Investigating Heart Rate, Heart Rate Variability, Experienced Effort and Likeability During Training of Three Types of Meditation," *International Journal of Psychophysiology* 97 (2015): 38–45.

12. Eileen Luders et al., "The Unique Brain Anatomy of Meditation Practitioners' Alterations in Cortical Gyrification," *Frontiers in Human Neuroscience* 6:34 (2012): 1–7.

13. The complexity of inferring that changes found are due to a given intervention—whether meditation or psychotherapy or a medicine rather than being "non-specific" effects of interventions in general—continues to be a crucial point in experiment design.

14. S. B. Goldberg et al., "Does the Five Facet Mindfulness Questionnaire Measure What We Think It Does? Construct Validity Evidence from an Active Controlled Randomized Clinical Trial," *Psychological Assessment* 28:8 (2016): 1009–14; doi:10.1037/pas0000233.

15. R. J. Davidson and Alfred W. Kazniak, "Conceptual and Methodological Issues in Research on Mindfulness and Meditation," *American Psychologist* 70:7 (2015): 581–92.

16. See also, e.g., Bhikkhu Bodhi, "What Does Mindfulness Really Mean? A Canonical Perspective," *Contemporary Buddhism* 12:1 (2011): 19–39; John Dunne, "Toward an Understanding of Non-Dual Mindfulness," *Contemporary Buddhism* 12:1 (2011) 71–88.

17. See, e.g, http://www.mindful.org/jon-kabat-zinn-defining-mindfulness/. Also J. Kabat-Zinn, "Mindfulness-Based Interventions in Context: Past, Present, and Future," *Clinical Psychology Science and Practice* 10 (2003): 145.

18. The Five Facet Mindfulness Questionnaire: R. A. Baer et al., "Using Self-Report Assessment Methods to Explore Facets of Mindfulness," *Assessment* 13 (2009): 27–45.

19. S. B. Goldberg et al., "The Secret Ingredient in Mindfulness Interventions? A Case for Practice Quality over Quantity," *Journal of Counseling Psychology* 61 (2014): 491–97.

20. J. Leigh et al., "Spirituality, Mindfulness, and Substance Abuse, *Addictive Behavior* 20:7 (2005): 1335–41.
21. E. Antonova et al., "More Meditation, Less Habituation: The Effect of Intensive Mindfulness Practice on the Acoustic Startle Reflex," *PLoS One* 10:5 (2015): 1–16; doi:10.1371/journal.pone.0123512.
22. D. B. Levinson et al., "A Mind You Can Count On: Validating Breath Counting as Behavioral Measure of Mindfulness," *Frontiers in Psychology* 5:1202 (2014); http://journal.frontiersin.org/Journal/110196/abstract.
23. Ibid.

CHAPTER FIVE. A MIND UNDISTURBED

1. St. Abba Dorotheus, quoted in E. Kadloubovsky and G. E. H. Palmer, *Early Fathers from the Philokalia* (London: Faber & Faber, 1971), p. 161.
2. Thomas Merton, "When the Shoe Fits," *The Way of Chuang Tzu* (New York: New Directions, 2010), p. 226.
3. Bruce S. McEwen, "Allostasis and Allostatic Load," *Neuropsychoparmacology* 22 (2000): 108–24.
4. Jon Kabat-Zinn, "Some Reflections on the Origins of MBSR, Skillful Means, and the Trouble with Maps," *Contemporary Buddhism* 12:1 (2011); doi:10.1080 /14639947.2011.564844.
5. Ibid.
6. Philippe R. Goldin and James J. Gross, "Effects of Mindfulness-Based Stress Reduction (MBSR) on Emotion Regulation in Social Anxiety Disorder," *Emotion* 10:1 (2010): 83–91; http://dx.doi.org/10.1037/a0018441.
7. Phillipe Goldin et al., "MBSR vs. Aerobic Exercise in Social Anxiety: fMRI of Emotion Regulation of Negative Self-Beliefs," *Social Cognitive and Affective Neuroscience Advance Access,* published August 27, 2012; doi:10.1093/scan/ nss054.
8. Alan Wallace, *The Attention Revolution: Unlocking the Power of the Focused Mind.* Somerville, MA: Wisdom Publications, 2006. For an exploration of the various meanings of "mindfulness," see B. Alan Wallace, "A Mindful Balance," *Tricycle* (Spring 2008): 60.
9. Gaelle Desbordes, "Effects of Mindful-Attention and Compassion Meditation Training on Amygdala Response to Emotional Stimuli in an Ordinary, Non-Meditative State," *Frontiers in Human Neuroscience* 6:292 (2012): 1–15; doi: 10.399/fnhum.2012.00292.
10. V. A. Taylor et al., "Impact of Mindfulness on the Neural Responses to Emotional Pictures in Experienced and Beginner Meditators," *NeuroImage* 57:4 (2011): 1524–1533; doi:10.1016/j.neuroimage.2011.06.001.
11. Tor D. Wager et al., "An fMRI-Based Neurologic Signature of Physical Pain," *NEJM* 368:15 (April 11, 2013): 1388–97.

12. See, e.g., James Austin, *Zen and the Brain: Toward an Understanding of Meditation and Consciousness* (Cambridge, MA: MIT Press, 1999).

13. Isshu Miura and Ruth Filler Sasaki, *The Zen Koan* (New York: Harcourt, Brace & World, 1965), p. xi.

14. Joshua A. Grant et al., "A Non-Elaborative Mental Stance and Decoupling of Executive and Pain-Related Cortices Predicts Low Pain Sensitivity in Zen Meditators," *Pain* 152 (2011): 150–56.

15. A. Golkar et al., "The Influence of Work-Related Chronic Stress on the Regulation of Emotion and on Functional Connectivity in the Brain," *PloS One* 9:9 (2014): e104550.

16. Stacey M. Schaefer et al., "Purpose in Life Predicts Better Emotional Recovery from Negative Stimuli," *PLoS One* 8:11 (2013): e80329; doi:10.1371/journal. pone.0080329.

17. Clifford Saron, "Training the Mind—The Shamatha Project," in A. Fraser, ed., *The Healing Power of Meditation* (Boston, MA: Shambhala Publications, 2013), pp. 45–65.

18. Baljinder K. Sahdra et al., "Enhanced Response Inhibition During Intensive Meditation Training Predicts Improvements in Self-Reported Adaptive Socioemotional Functioning," *Emotion* 11:2 (2011): 299–312.

19. Margaret E. Kemeny et al., "Contemplative/Emotion Training Reduces Negative Emotional Behavior and Promotes Prosocial Responses," *Emotion* 1:2 (2012): 338.

20. Melissa A. Rosenkranz et al., "Reduced Stress and Inflammatory Responsiveness in Experienced Meditators Compared to a Matched Healthy Control Group," *Psychoneuroimmunology* 68 (2016): 117–25. The long-term meditators all had practiced vipassana and loving-kindness meditation over a period of at least three years, did daily practice of at least thirty minutes, and had also done several intensive meditation retreats. Each was matched on age and sex with a nonmeditating volunteer to create a comparison group. They also gave saliva samples at several points in the experiment, which revealed their levels of cortisol. There was no active control group here, for two reasons. When the measures used are biological rather than self-report, the outcomes are far less susceptible to bias. And, as with Cliff's three-month course, it would be impossible to create an active control akin to 9,000 hours of meditation over three years or more.

21. T. R. A. Kral et al., "Meditation Training Is Associated with Altered Amygdala Reactivity to Emotional Stimuli," under review, 2017.

22. If Richie were analyzing the data in the same way as that used in most other studies, none of these differences would have emerged. The peak of the amygdala response was identical in these groups. However, the response of the meditators who had practiced the longest showed the fastest recovery. This may be a neural echo of "nonstickiness"—showing an appropriate initial response to a disturbing image, but then not having that response linger.

CHAPTER SIX. PRIMED FOR LOVE

1. The Desert Fathers were early Christian hermits who lived in communities located in remote areas of Egypt's desert in the early centuries AD. There they could better focus on their religious practices, mainly the recitation of *Kyrie Eleison* (a Greek phrase meaning "Lord, have mercy"), a Christian "mantra." These hermit communities were the historic predecessors of Christian orders for monks and nuns; repetition of *Kyrie Eleison* remains a primary practice among Eastern Orthodox monks, e.g., those on Mount Athos. Historical records suggest that Christian monks from Egypt settled on Mount Athos in the seventh century, fleeing Islamic conquest. Helen Waddell, *The Desert Fathers* (Ann Arbor: University of Michigan Press, 1957).

2. The Good Samaritan setup was an experiment, one of an extensive, systematic series of studies of the conditions that encourage or inhibit altruistic acts. Daniel Batson, *Altruism in Humans* (New York: Oxford University Press, 2011).

3. Sharon Salzberg, *Lovingkindness: The Revolutionary Art of Happiness* (Boston: Shambhala, 2002).

4. Arnold Kotler, ed., *Worlds in Harmony: Dialogues on Compassionate Action* (Berkeley: Parallax Press, 1992).

5. The researchers note that self-criticism is not limited to depression; it shows up in a range of emotional problems. Like these researchers, we'd like to see a study that shows a meditation-induced rise in self-compassion coupled with a similar shift in related brain circuitry. See Ben Shahar, "A Wait-List Randomized Controlled Trial of Loving-Kindness Meditation Programme for Self-Criticism," *Clinical Psychology and Psychotherapy* (2014); doi:10.1002/cpp.1893.

6. See, e.g., Jean Decety, "The Neurodevelopment of Empathy," *Developmental Neuroscience* 32 (2010): 257–67.

7. Olga Klimecki et al., "Functional Neural Plasticity and Associated Changes in Positive Affect after Compassion Training," *Cerebral Cortex* 23:7 (July 2013) 1552–61.

8. Olga Klimecki et al., "Differential Pattern of Functional Brain Plasticity after Compassion and Empathy Training," *Social Cognitive and Affective Neuroscience* 9:6 (June 2014): 873–79; doi:10.1093/scan/nst060.

9. Thich Nhat Hanh, "The Fullness of Emptiness," *Lion's Roar*, August 6, 2012. "Kuan" is sometimes rendered as "Kwan," "Guan," or "Quan."

10. Gaelle Desbordes, "Effects of Mindful-Attention and Compassion Meditation Training on Amygdala Response to Emotional Stimuli in an Ordinary, Non-Meditative State," *Frontiers in Human Neuroscience* 6:292 (2012): 1–15; doi: 10.399/fnhum.2012.00292.

11. Cendri A. Hutcherson et al., "Loving-Kindness Meditation Increases Social Connectedness," *Emotion* 8:5 (2008): 720–24.

12. Helen Y. Weng et al., "Compassion Training Alters Altruism and Neural Responses to Suffering," *Psychological Science*, published online May 21, 2013; http://pss.sagepub.com/content/early/2013/05/20/0956797612469537.

13. Julieta Galante, "Loving-Kindness Meditation Effects on Well-Being and Altruism: A Mixed-Methods Online RCT," *Applied Psychology: Health and Well-Being* (2016); doi:10.1111/aphw.12074.

14. Antoine Lutz et al., "Regulation of the Neural Circuitry of Emotion by Compassion Meditation: Effects of Meditative Expertise," *PLoS One* 3:3 (2008): e1897; doi:10.1371/journal.pone.0001897.

15. J. A. Brefczynski-Lewis et al., "Neural Correlates of Attentional Expertise in Long-Term Meditation Practitioners," *Proceedings of the National Academy of Sciences* 104:27 (2007): 11483–88.

16. Clifford Saron, presentation at the Second International Conference on Contemplative Science, San Diego, November 2016.

17. Abigail A. Marsh et al., "Neural and Cognitive Characteristics of Extraordinary Altruists," *Proceedings of the National Academy of Sciences* 111:42 (2014), 15036–41; doi: 10.1073/pnas.1408440111.

18. There are numerous factors at work in altruism, but the ability to feel the suffering of someone else seems a key ingredient. To be sure, the changes in the meditators were not as strong nor as long-lasting as the structural brain patterns unique to the kidney donors. See Desbordes, "Effects of Mindful-Attention and Compassion Meditation Training on Amygdala Response to Emotional Stimuli in an Ordinary, Non-Meditative State," 2012.

19. Tania Singer and Olga Klimecki, "Empathy and Compassion," *Current Biology* 24:15 (2014): R875–R878.

20. Weng et al., "Compassion Training Alters Altruism and Neural Responses to Suffering," 2013.

21. Tania Singer et al., "Empathy for Pain Involves the Affective but Not Sensory Components of Pain," *Science* 303:5661 (2004): 1157–62; doi:10.1126/science.1093535.

22. Klimecki et al., "Functional Neural Plasticity and Associated Changes in Positive Affect after Compassion Training."

23. Bethany E. Kok and Tania Singer, "Phenomenological Fingerprints of Four Meditations: Differential State Changes in Affect, Mind-Wandering, Meta-Cognition, and Interoception Before and After Daily Practice Across 9 Months of Training," *Mindfulness*, published online August 19, 2016; doi: 10.1007/s12671-016-0594-9.

24. Yoni Ashar et al., "Effects of Compassion Meditation on a Psychological Model of Charitable Donation," *Emotion*, published online March 28, 2016, http:///dx.doi.org/10.1037/emo0000119.

25. Paul Condon et al., "Meditation Increases Compassionate Response to Suffering," *Psychological Science* 24:10 (August 2013): 1171–80; doi:10.1177/0956797613485603.

26. Desbordes et al., "Effects of Mindful-Attention and Compassion Meditation Training on Amygdala Response to Emotional Stimuli in an Ordinary, Non-Meditative State," 2012. Both groups practiced for a total of at least twenty hours. All the volunteers had brain scanning before and after the training; the second group were scanned while they were simply at rest, not meditating.

27. See, for example, Derntl et al., "Multidimensional Assessment of Empathic Abilities: Neural Correlates and Gender Differences," *Psychoneuroimmunology* 35 (2010): 67–82.

28. L. Christov-Moore et al., "Empathy: Gender Effects in Brain and Behavior," *Neuroscience & Biobehavioral Reviews* 4:46 (2014): 604–27; doi:10.1016/j.neubiorev.2014.09.001.Empathy.

29. M. P. Espinosa and J. Kovářík, "Prosocial Behavior and Gender," *Frontiers in Behavioral Neuroscience* 9 (2015): 1–9; doi:10.3389/fnbeh.2015.00088.

30. The Dalai Lama extends this sentiment infinitely. Although we have no proof, there may well be other worlds in galaxies near or far with their own life-forms. If so, he assumes they, too, would wish to avoid suffering and want happiness.

31. A. J. Greenwald and M. R. Banaji, "Implicit Social Cognition: Attitudes, Self-Esteem, and Stereotypes," *Psychological Review* 102:1 (1995): 4–27; doi:10.1037/0033-295X.102.1.4.

32. Y. Kang et al., "The Nondiscriminating Heart: Lovingkindness Meditation Training Decreases Implicit Intergroup Bias," *Journal of Experimental Psychology* 143:3 (2014): 1306–13; doi:10.1037/a0034150.

33. The Dalai Lama made these remarks in Dunedin, New Zealand, on June 10, 2013, as recorded by Jeremy Russell at www.dalailama.org.

CHAPTER SEVEN. ATTENTION!

1. Charlotte Joko Beck, *Nothing Special: Living Zen* (New York: HarperCollins, 1993), p. 168.

2. Akira Kasamatsu and Tomio Hirai, "An Electroencephalographic Study on Zen Meditation (Zazen)," *Psychiatry and Clinical Neurosciences* 20:4 (1966): 325–36.

3. Elena Antonova et al., "More Meditation, Less Habituation: The Effect of Intensive Mindfulness Practice on the Acoustic Startle Reflex," *PLoS One* 10:5 (2015): 1–16; doi:10.1371/journal.pone.0123512. The meditators were instructed to stay in "open awareness" during the noises, and the meditation-naive controls were instructed to "remain alert and awake throughout the experiment . . . and to return their awareness to the surroundings if they caught themselves mind-wandering."

4. T. R. A. Kral et al., "Meditation Training Is Associated with Altered Amygdala Reactivity to Emotional Stimuli," under review, 2017.

5. Amishi Jha et al., "Mindfulness Training Modifies Subsystems of Attention," *Cognitive, Affective, & Behavioral Neuroscience* 7:2 (2007): 109–19; http://www.ncbi.nlm.nih.gov/pubmed/17672382.

6. Catherine E. Kerr et al., "Effects of Mindfulness Meditation Training on Anticipatory Alpha Modulation in Primary Somatosensory Cortex," *Brain Research Bulletin* 85 (2011): 98–103.

7. Antoine Lutz et al., "Mental Training Enhances Attentional Stability: Neural and Behavioral Evidence," *Journal of Neuroscience* 29:42 (2009): 13418–27; Heleen A. Slagter et al., "Theta Phase Synchrony and Conscious Target Perception: Impact of Intensive Mental Training," *Journal of Cognitive Neuroscience* 21:8 (2009): 1536–49. An active control group, who were taught mindfulness during a one-hour session at the start and end of the three-month period and were instructed to practice for twenty minutes per day, did no better after that training than before.

8. Katherine A. MacLean et al., "Intensive Meditation Training Improves Perceptual Discrimination and Sustained Attention," *Psychological Science* 21:6 (2010): 829–39.

9. H. A. Slagter et al., "Mental Training Affects Distribution of Limited Brain Resources," *PLoS Biology* 5:6 (2007): e138; doi:10.1371/journal.pbio.0050138. Among nonmeditating controls tested at the same intervals, there was no change in the attentional blink.

10. Sara van Leeuwen et al., "Age Effects on Attentional Blink Performance in Meditation," *Consciousness and Cognition,* 18 (2009): 593–99.

11. Lorenzo S. Colzato et al., "Meditation-Induced States Predict Attentional Control over Time," *Consciousness and Cognition* 37 (2015): 57–62.

12. E. Ophir et al., "Cognitive Control in Multi-Taskers," *Proceedings of the National Academy of Sciences* 106:37 (2009): 15583–87.

13. Clifford Nass, in an NPR interview, as quoted in *Fast Company,* February 2, 2014.

14. Thomas E. Gorman and C. Shawn Gree, "Short-Term Mindfulness Intervention Reduces the Negative Attentional Effects Associated with Heavy Media Multitasking," *Scientific Reports* 6 (2016): 24542; doi:10.1038/srep24542.

15. Michael D. Mrazek et al., "Mindfulness and Mind Wandering: Finding Convergence through Opposing Constructs," *Emotion* 12:3 (2012): 442–48.

16. Michael D. Mrazek et al., "Mindfulness Training Improves Working Memory Capacity and GRE Performance While Reducing Mind Wandering," *Psychological Science* 24:5 (2013): 776–81.

17. Bajinder K. Sahdra et al., "Enhanced Response Inhibition During Intensive Meditation Predicts Improvements in Self-Reported Adaptive Socioemotional Functioning," *Emotion* 11:2 (2011): 299–312.

18. Sam Harris, *Waking Up: A Guide to Spirituality Without Religion* (NY: Simon & Schuster, 2015), p. 144.

19. See, e.g., Daniel Kahneman, *Thinking, Fast and Slow* (New York: Farrar, Straus and Giroux, 2011).

20. R. C. Lapate et al., "Awareness of Emotional Stimuli Determines the Behavioral Consequences of Amygdala Activation and Amygdala-Prefrontal Connectivity," *Scientific Reports* 20:6 (2016): 25826; doi:10.1038/srep25826.

21. Benjamin Baird et al., "Domain-Specific Enhancement of Metacognitive Ability Following Meditation Training," *Journal of Experimental Psychology: General* 143:5 (2014): 1972–79; http://dx.doi.org/10.1037/a0036882. Both the mindfulness and active control groups took forty-five-minute classes four times a week for two weeks, along with home practice for fifteen minutes daily.

22. Amishi Jha et al., "Mindfulness Training Modifies Subsystems of Attention," *Cognitive Affective and Behavioral Neuroscience* 7:2 (2007): 109–19; doi: 10.3758/cabn.7.2.109.

CHAPTER EIGHT. LIGHTNESS OF BEING

1. Marcus Raichle et al., "A Default Mode of Brain Function," *Proceedings of the National Academy of Sciences* 98 (2001): 676–82.

2. M. F. Mason et al., "Wandering Minds: The Default Network and Stimulus-Independent Thought," *Science* 315:581 (2007): 393–95; doi:10.1126/science .1131295.

3. Judson Brewer et al., "Meditation Experience Is Associated with Differences in Default Mode Network Activity and Connectivity," *Proceedings of the National Academy of Sciences* 108:50 (2011): 1–6; doi:10.1073/pnas.1112029108.

4. Fakhruddin Iraqi, a thirteenth-century Sufi poet, quoted in James Fadiman and Robert Frager, *Essential Sufism* (New York: HarperCollins, 1997).

5. Abu Said of Mineh, quoted in P. Rice, *The Persian Sufis* (London: Allen & Unwin, 1964), p. 34.

6. David Creswell et al., "Alterations in Resting-State Functional Connectivity Link Mindfulness Meditation with Reduced Interleukin-6: A Randomized Controlled Trial," *Biological Psychiatry* 80 (2016): 53–61.

7. Brewer et al., "Meditation Experience Is Associated with Differences in Default Mode Network Activity and Connectivity."

8. Kathleen A. Garrison et al., "BOLD Signals and Functional Connectivity Associated with Loving Kindness Meditation," *Brain and Behavior* 4:3 (2014): 337–47.

9. Aviva Berkovich-Ohana et al., "Alterations in Task-Induced Activity and Resting-State Fluctuations in Visual and DMN Areas Revealed in Long-Term Meditators," *NeuroImage* 135 (2016): 125–34.

10. Giuseppe Pagnoni, "Dynamical Properties of BOLD Activity from the Ventral Posteromedial Cortex Associated with Meditation and Attentional Skills," *Journal of Neuroscience* 32:15 (2012): 5242–49.

11. V. A. Taylor et al., "Impact of Meditation Training on the Default Mode Network during a Restful State," *Social Cognitive and Affective Neuroscience* 8 (2013): 4–14.

12. D. B. Levinson et al., "A Mind You Can Count On: Validating Breath Counting as a Behavioral Measure of Mindfulness," *Frontiers in Psychology* 5 (2014); http://journal.frontiersin.org/Journal/110196/abstract.

13. Cole Koparnay, Center for Healthy Minds, University of Wisconsin, in preparation. This study applied stricter criteria for brain changes than earlier ones that have reported various increases in meditator brain volume.

14. Then again, perhaps a subset of meditators goes down a path that makes them more aloof and cold or indifferent. Offsetting this tendency might be one reason so many traditions emphasize compassion and devotion, which are "juicy."

15. Arthur Zajonc, personal communication.

16. Kathleen Garrison et al., "Effortless Awareness: Using Real Time Neurofeedback to Investigate Correlates of Posterior Cingulate Cortex Activity in Meditators' Self-Report," *Frontiers in Human Neuroscience* 7:440 (August 2013): 1–9.

17. Anna-Lena Lumma et al., "Is Meditation Always Relaxing? Investigating Heart Rate, Heart Rate Variability, Experienced Effort and Likeability During Training of Three Types of Meditation," *International Journal of Psychophysiology* 97 (2015): 38–45.

18. See Daniel Goleman, *Destructive Emotions: How Can We Overcome Them?* (New York: Bantam, 2003).

CHAPTER NINE. MIND, BODY, AND GENOME

1. Natalie A. Morone et al., "A Mind-Body Program for Older Adults with Chronic Low Back Pain: A Randomized Trial," *JAMA Internal Medicine* 176:3 (2016): 329–37.

2. M. M. Veehof, "Acceptance- and Mindfulness-Based Interventions for the Treatment of Chronic Pain: A Meta-Analytic Review, 2016," *Cognitive Behaviour Therapy* 45:1 (2016): 5–31.

3. Paul Grossman et al., "Mindfulness-Based Intervention Does Not Influence Cardiac Autonomic Control or Pattern of Physical Activity in Fibromyalgia in Daily Life: An Ambulatory, Multi-Measure Randomized Controlled Trial," *Clinical Journal of Pain* (2017); doi: 10.1097/AJP.0000000000000420.

4. Elizabeth Cash et al., "Mindfulness Meditation Alleviates Fribromyalgia Symptoms in Women: Results of a Randomized Clinical Trial," *Annals of Behavioral Medicine* 49:3 (2015): 319–30.

5. Melissa A. Rosenkranz et al., "A Comparison of Mindfulness-Based Stress Reduction and an Active Control in Modulation of Neurogenic Inflammation," *Brain, Behavior, and Immunity* 27 (2013): 174–84.

6. Melissa A. Rosenkranz et al., "Neural Circuitry Underlying the Interaction Between Emotion and Asthma Symptom Exacerbation," *Proceedings of the National Academy of Sciences* 102:37 (2005): 13319–24; http://doi.org/10.1073/pnas.0504365102.

7. Jon Kabat-Zinn et al., "Influence of a Mindfulness Meditation-Based Stress Reduction Intervention on Rates of Skin Clearing in Patients with Moderate to Severe Psoriasis Undergoing Phototherapy (UVB) and Photochemotherapy (PUVA)," *Psychosomatic Medicine* 60 (1988): 625–32.

8. Melissa A. Rosenkranz et al., "Reduced Stress and Inflammatory Responsiveness in Experienced Meditators Compared to a Matched Healthy Control Group," *Psychoneuroimmunology* 68 (2016): 117–25.

9. E. Walsh, "Brief Mindfulness Training Reduces Salivary IL-6 and TNF-α in Young Women with Depressive Symptomatology," *Journal of Consulting and Clinical Psychology* 84:10 (2016): 887–97; doi:10.1037/ccp0000122; T. W. Pace et al., "Effect of Compassion Meditation on Neuroendocrine, Innate Immune and Behavioral Responses to Psychological Stress," *Psychoneuroimmunology* 34 (2009): 87–98.

10. David Creswell et al., "Alterations in Resting-State Functional Connectivity Link Mindfulness Meditation with Reduced Interleukin-6: A Randomized Controlled Trial," *Biological Psychiatry* 80 (2016): 53–61.

11. Daniel Goleman, "Hypertension? Relax," *New York Times Magazine,* December 11, 1988.

12. Jeanie Park et al., "Mindfulness Meditation Lowers Muscle Sympathetic Nerve Activity and Blood Pressure in African-American Males with Chronic Kidney Disease," *American Journal of Physiology—Regulatory, Integrative and Comparative Physiology* 307:1 (July 1, 2014), R93–R101; published online May 14, 2014; doi:10.1152/ajpregu.00558.2013.

13. John O. Younge, "Mind-Body Practices for Patients with Cardiac Disease: A Systematic Review and Meta-Analysis," *European Journal of Preventive Cardiology* 22:11 (2015): 1385–98.

14. Perla Kaliman et al., "Rapid Changes in Histone Deacetylases and Inflammatory Gene Expression in Expert Meditators," *Psychoneuroendocrinology* 40 (2014): 96–107.

15. J. D. Creswell et al., "Mindfulness-Based Stress Reduction Training Reduces Loneliness and Pro-Inflammatory Gene Expression in Older Adults: A Small Randomized Controlled Trial," *Brain, Behavior, and Immunity* 26 (2012): 1095–1101.

16. J. A. Dusek, "Genomic Counter-Stress Changes Induced by the Relaxation Response," *PLoS One* 3:7 (2008): e2576; M. K. Bhasin et al., "Relaxation Response Induces Temporal Transcriptome Changes in Energy Metabolism, Insulin Secretion and Inflammatory Pathways," *PLoS One* 8.5 (2013): e62817.

17. H. Lavretsky et al., "A Pilot Study of Yogic Meditation for Family Dementia Caregivers with Depressive Symptoms: Effects on Mental Health, Cognition, and Telomerase Activity," *International Journal of Geriatric Psychiatry* 28:1 (2013): 57–65.

18. N. S. Schutte and J. M. Malouff, "A Meta-Analytic Review of the Effects of Mindfulness Meditation on Telomerase Activity," *Psychoneuroendocrinology* 42 (2014): 45–48; http://doi.org/10.1016/j.psyneuen.2013.12.017.

19. Tonya L. Jacobs et al., "Intensive Meditation Training, Immune Cell Telomerase Activity, and Psychological Mediators," *Psychoneuroendocrinology* 36:5 (2011): 664–81; http://doi.org/10.1016/j.psyneuen.2010.09.010.

20. Elizabeth A. Hoge et al., "Loving-Kindness Meditation Practice Associated with Longer Telomeres in Women," *Brain, Behavior, and Immunity* 32 (2013): 159–63.

21. Christine Tara Peterson et al., "Identification of Altered Metabolomics Profiles Following a *Panchakarma*-Based Ayurvedic Intervention in Healthy Subjects: The Self-Directed Biological Transformation Initiative (SBTI)," *Nature: Scientific Reports* 6 (2016): 32609; doi:10.1038/srep32609.

22. A. L. Lumma et al., "Is Meditation Always Relaxing? Investigating Heart Rate, Heart Rate Variability, Experienced Effort and Likeability During Training of Three Types of Meditation," *International Journal of Psychophysiology* 97:1 (2015): 38–45.

23. Antoine Lutz et al., "BOLD Signal in Insula Is Differentially Related to Cardiac Function during Compassion Meditation in Experts vs. Novices," *NeuroImage* 47:3 (2009): 1038–46; http://doi.org/10.1016/j.neuroimage.2009.04.081.

24. J. Wielgosz et al., "Long-Term Mindfulness Training Is Associated with Reliable Differences in Resting Respiration Rate," *Scientific Reports* 6 (2016): 27533; doi:10.1038/srep27533.

25. Sara Lazar et al., "Meditation Experience Is Associated with Increased Cortical Thickness," *Neuroreport* 16 (2005): 1893–97. The study compared twenty vipassana practitioners (average around 3,000 hours lifetime experience) with age- and gender-matched controls.

26. Kieran C. R. Fox, "Is Meditation Associated with Altered Brain Structure? A Systematic Review and Meta-Analysis of Morphometric Neuroimaging in Meditation Practitioners," *Neuroscience and Biobehavioral Reviews* 43 (2014): 48–73.

27. Eileen Luders et al., "Estimating Brain Age Using High-Resolution Pattern Recognition: Younger Brains in Long-Term Meditation Practitioners," *NeuroImage* (2016); doi:10.1016/j.neuroimage. 2016.04.007.

28. Eileen Luders et al., "The Unique Brain Anatomy of Meditation Practitioners' Alterations in Cortical Gyrification," *Frontiers in Human Neuroscience* 6:34 (2012): 1–7.

29. For example, B. K. Holzel et al., "Mindfulness Meditation Leads to Increase in Regional Grey Matter Density," *Psychiatry Research: Neuroimaging* 191 (2011): 36–43.

30. S. Coronado-Montoya et al., "Reporting of Positive Results in Randomized Controlled Trials of Mindfulness-Based Mental Health Interventions," *PLoS One* 11:4 (2016): e0153220; http://doi.org/10.1371/journal.pone.0153220.

31. Cole Korponay, in preparation.

32. A. Tusche et al., "Decoding the Charitable Brain: Empathy, Perspective Taking, and Attention Shifts Differentially Predict Altruistic Giving," *Journal of Neuroscience* 36:17 (2016):4719–32. doi:10.1523/JNEUROSCI.3392-15.2016.

33. S. K. Sutton and R. J. Davidson, "Prefrontal Brain Asymmetry: A Biological Substrate of the Behavioral Approach and Inhibition Systems," *Psychological Science* 8:3 (1997): 204–10; http://doi.org/10.1111/j.1467-9280.1997.tb00413.x.

34. Daniel Goleman, *Destructive Emotions: How Can We Overcome Them?* (New York: Bantam, 2003).

35. P. M. Keune et al., "Mindfulness-Based Cognitive Therapy (MBCT), Cognitive Style, and the Temporal Dynamics of Frontal EEG Alpha Asymmetry in Recurrently Depressed Patients," *Biological Psychology* 88:2–3 (2011): 243–52; http://doi.org/10.1016/j.biopsycho.2011.08.008.

36. P. M. Keune et al., "Approaching Dysphoric Mood: State-Effects of Mindfulness Meditation on Frontal Brain Asymmetry," *Biological Psychology* 93:1 (2013): 105–13; http://doi.org/10.1016/j.biopsycho.2013.01.016.

37. E. S. Epel et al., "Meditation and Vacation Effects Have an Impact on Disease-Associated Molecular Phenotypes," *Nature* 6 (2016): e880; doi:10.1038/tp.2016.164.

38. The Stephen E. Straus Distinguished Lecture in the Science of Complementary Health Therapies.

CHAPTER TEN. MEDITATION AS PSYCHOTHERAPY

1. Tara Bennett-Goleman, *Emotional Alchemy: How the Mind Can Heal the Heart* (New York: Harmony Books, 2001).

2. Zindel Segal, Mark Williams, John Teasdale, et al., *Mindfulness-Based Cognitive Therapy for Depression* (New York: Guilford Press, 2003); John Teasdale et al., "Prevention of Relapse/Recurrence in Major Depression by Mindfulness-Based Cognitive Therapy," *Journal of Consulting and Clinical Psychology* 68:4 (2000): 615–23.

3. Madhav Goyal et al., "Meditation Programs for Psychological Stress and Well-Being: A Systematic Review and Meta-Analysis," *JAMA Internal Medicine*, published online January 6, 2014; doi:10.1001/jamainternmed.2013.13018.

4. J. Mark Williams et al., "Mindfulness-Based Cognitive Therapy for Preventing Relapse in Recurrent Depression: A Randomized Dismantling Trial," *Journal of Consulting and Clinical Psychology* 82:2 (2014): 275–86.

5. Alberto Chiesa, "Mindfulness-Based Cognitive Therapy vs. Psycho-Education for Patients with Major Depression Who Did Not Achieve Remission Following Anti-Depressant Treatment," *Psychiatry Research* 226 (2015): 174–83.

6. William Kuyken et al., "Efficacy of Mindfulness-Based Cognitive Therapy in Prevention of Depressive Relapse," *JAMA Psychiatry* (April 27, 2016); doi:10.1001/jamapsychiatry.2016.0076.

7. Zindel Segal, presentation at the International Conference on Contemplative Science, San Diego, November 18–20, 2016.

8. Sona Dimidjian et al., "Staying Well During Pregnancy and the Postpartum: A Pilot Randomized Trial of Mindfulness-Based Cognitive Therapy for the Prevention of Depressive Relapse/Recurrence," *Journal of Consulting and Clinical Psychology* 84:2 (2016): 134–45.

9. S. Nidich et al., "Reduced Trauma Symptoms and Perceived Stress in Male Prison Inmates through the Transcendental Meditation Program: A Randomized Controlled Trial," *Permanente Journal* 20:4 (2016): 43–47; http://doi.org /10.7812/TPP/16-007.

10. Filip Raes et al., "School-Based Prevention and Reduction of Depression in Adolescents: A Cluster-Randomized Controlled Trial of a Mindfulness Group," *Mindfulness*, March 2013; doi:10.1007/s12671-013-0202-1.

11. Philippe R. Goldin and James J. Gross, "Effects of Mindfulness-Based Stress Reduction (MBSR) on Emotion Regulation in Social Anxiety Disorder," *Emotion* 10:1 (2010): 83–91; http://dx.doi.org/10.1037/a0018441.

12. David J. Kearney et al., "Loving-Kindness Meditation for Post-Traumatic Stress Disorder: A Pilot Study," *Journal of Traumatic Stress* 26 (2013): 426–34. The VA researchers point out that their promising results call for a follow-up study, which as of this writing is in progress. This follow-up study has 130 veterans with PTSD, randomized into an active control group, and a four-year timeline. Loving-kindness meditation is being compared to what's considered a "gold standard" treatment for PTSD, a variety of cognitive therapy, in the active control. The hypothesis: loving-kindness will work as well, but via different mechanisms.

13. Another anecdotal report: P. Gilbert and S. Procter, "Compassionate Mind Training for People with High Shame and Self-Criticism: Overview and Pilot Study of a Group Therapy Approach," *Clinical Psychology & Psychotherapy* 13 (2006): 353–79.

14. Jay Michaelson, *Evolving Dharma: Meditation, Buddhism, and the Next Generation of Enlightenment* (Berkeley: Evolver Publications, 2013). In popular use the phrase "dark night" in a spiritual journey has become twisted a bit in meaning from its original sense. The seventeenth-century Spanish mystic St. John of the Cross famously first used the term—but to describe the mysterious ascent through an unknown territory to an ecstatic merger with the divine. Today, though, "dark night" means miring in the fears and such that threatening our worldly identity can bring.

15. Daniel Goleman, "Meditation as Meta-Therapy: Hypotheses Toward a Proposed Fifth State of Consciousness," *Journal of Transpersonal Psychology* 3:1 (1971): 1–26.

16. Jack Kornfield, *The Wise Heart: A Guide to the Universal Teachings of Buddhist Psychology* (New York: Bantam, 2009).

17. Daniel Goleman and Mark Epstein, "Meditation and Well-Being: An Eastern Model of Psychological Health," *ReVision* 3:2 (1980): 73–84. Reprinted in Roger Walsh and Deane Shapiro, *Beyond Health and Normality* (New York: Van Nostrand Reinhold, 1983).

18. *Thoughts Without a Thinker: Psychotherapy from a Buddhist Perspective* (New York: Basic Books, 1995) was Mark Epstein's first book; *Advice Not Given: A Guide to Getting over Yourself* (New York: Penguin Press, 2018) will be his next.

CHAPTER ELEVEN. A YOGI'S BRAIN

1. François Jacob discovered that enzyme expression levels in cells occur through the mechanisms of DNA transcription. For this discovery he won a Nobel Prize in 1965.

2. For several years Matthieu was a board member of the Mind and Life Institute, and has long engaged with the scientists connected to that community as well as in many scientific dialogues with the Dalai Lama.

3. Antoine Lutz et al., "Long-Term Meditators Self-Induce High-Amplitude Gamma Synchrony During Mental Practice," *Proceedings of the National Academy of Sciences* 101:46 (2004): 16369; http://www.pnas.org/content/101/46/16369 .short.

4. Dilgo Khyentse Rinpoche (1910–1991).

5. Lawrence K. Altman, *Who Goes First?* (New York: Random House, 1987).

6. Francisco J. Varela and Jonathan Shear, "First-Person Methodologies: What, Why, How?" *Journal of Consciousness Studies* 6:2–3 (1999): 1–14.

7. H. A. Slagter et al., "Mental Training as a Tool in the Neuroscientific Study of Brain and Cognitive Plasticity," *Frontiers in Human Neuroscience* 5:17 (2011); doi:10.3389/fnhum.2011.00017.

8. The curriculum has been developed by the Tibet-Emory Science Project, under the codirection of Geshe Lobsang Tenzin Negi. To celebrate the new curriculum, Richie was part of a meeting with the Dalai Lama, scientists, philosophers, and contemplatives at the Drepung Monastery, a Tibetan Buddhist outpost in the South Indian state of Karnataka. Mind and Life XXVI, "Mind, Brain, and Matter: A Critical Conversation between Buddhist Thought and Science," Mundgod, India, 2013.

9. At the time John Dunne was an assistant professor in the Department of Languages and Cultures of Asia at the University of Wisconsin; now he holds a chair as the Distinguished Professor of Contemplative Humanities, affiliated with Richie's research program there.

10. Antoine Lutz et al., "Long-Term Meditators Self-Induce High-Amplitude Gamma Synchrony During Mental Practice," *Proceedings of the National Academy of Sciences* 101:46 (2004): 16369. http://www.pnas.org/content/101/46/ 16369.short.

11. Tulku Urgyen's father, in turn, is said to have done more than thirty years of retreat over the course of his lifetime. And Tulku Urgyen's great-grandfather the legendary Chokling Rinpoche was a spiritual giant who founded a still-vibrant practice lineage. See Tulku Urgyen, trans. Erik Pema Kunzang, *Blazing Splendor* (Kathmandu: Blazing Splendor Publications, 2005).

CHAPTER TWELVE. HIDDEN TREASURE

1. Third Dzogchen Rinpoche, trans. Cortland Dahl, *Great Perfection, Volume Two: Separation and Breakthrough* (Ithaca, NY: Snow Lion Publications, 2008), p. 181.
2. F. Ferrarelli et al., "Experienced Mindfulness Meditators Exhibit Higher Parietal-Occipital EEG Gamma Activity during NREM Sleep," *PLoS One* 8:8 (2013): e73417; doi:10.1371/journal.pone.0073417. This fits what yogis report, and we strongly suspect we would find it in them, too (that study of sleep in Tibetan yogis has not yet been done—although they actually do a practice to cultivate meditative awareness during sleep).
3. Antoine Lutz et al., "Long-Term Meditators Self-Induce High-Amplitude Gamma Synchrony During Mental Practice," *Proceedings of the National Academy of Sciences* 101:46 (2004): 16369; http://www.pnas.org/content/101/46/16369.short.
4. Antoine Lutz et al., "Regulation of the Neural Circuitry of Emotion by Compassion Meditation: Effects of Meditative Expertise," *PLoS One* 3:3 (2008): e1897; doi:10.1371/journal.pone.0001897.
5. For the week leading up to their brain scan session, the novices spent twenty minutes a day generating this state of positivity toward all.
6. Lutz et al., "Regulation of the Neural Circuitry of Emotion by Compassion Meditation: Effects of Meditative Expertise."
7. Judson Brewer et al., "Meditation Experience Is Associated with Differences in Default Mode Network Activity and Connectivity," *Proceedings of the National Academy of Sciences* 108:50 (2011): 1–6; doi:10.1073/pnas.1112029108.
8. https://www.freebuddhistaudio.com/texts/meditation/Dilgo_Khyentse_Rinpoche/FBA13_Dilgo_Khyentse_Rinpoche_on_Maha_Ati.pdf.
9. The Third Khamtrul Rinpoche, trans. Gerardo Abboud, *The Royal Seal of Mahamudra* (Boston: Shambhala, 2014), p. 128.
10. Anna-Lena Lumma et al., "Is Meditation Always Relaxing? Investigating Heart Rate, Heart Rate Variability, Experienced Effort and Likeability During Training of Three Types of Meditation," *International Journal of Psychophysiology* 97 (2015): 38–45.
11. R. van Lutterveld et al., "Source-Space EEG Neurofeedback Links Subjective Experience with Brain Activity during Effortless Awareness Meditation," *NeuroImage* (2016); doi:10.1016/j.neuroimage.2016.02.047.

12. K. A. Garrison et al., "Effortless Awareness: Using Real Time Neurofeedback to Investigate Correlates of Posterior Cingulate Cortex Activity in Meditators' Self-Report," *Frontiers in Human Neuroscience* 7 (August 2013): 1–9; doi:10.3389/fnhum.2013.00440.

13. Antoine Lutz et al., "BOLD Signal in Insula Is Differentially Related to Cardiac Function during Compassion Meditation in Experts vs. Novices," *NeuroImage* 47:3 (2009): 1038–46; http://doi.org/10.1016/j.neuroimage.2009.04.081.

CHAPTER THIRTEEN. ALTERING TRAITS

1. Milarepa in Matthieu Ricard, *On the Path to Enlightenment* (Boston: Shambhala, 2013), p. 122.

2. Judson Brewer et al., "Meditation Experience Is Associated with Differences in Default Mode Network Activity and Connectivity," *Proceedings of the National Academy of Sciences* 108:50 (2011): 1–6; doi:10.1073/pnas.1112029108. V. A. Taylor et al., "Impact of Mindfulness on the Neural Responses to Emotional Pictures in Experienced and Beginner Meditators," *NeuroImage* 57:4 (2011): 1524–33; doi:101016/j.neuroimage.2011.06.001.

3. Francis de Sales, quoted in Aldous Huxley, *The Perennial Philosophy* (New York: Harper & Row, 1947), p. 285.

4. Wendy Hasenkamp and her team used fMRI to identify the brain regions engaged by each of these steps. Wendy Hasenkamp et al., "Mind Wandering and Attention during Focused Meditation: A Fine-Grained Temporal Analysis during Fluctuating Cognitive States," *NeuroImage* 59:1 (2012): 750–60; Wendy Hasenkamp and L. W. Barsalou, "Effects of Meditation Experience on Functional Connectivity of Distributed Brain Networks," *Frontiers in Human Neuroscience* 6:38 (2012); doi:10.3389/fnhum.2012.00038.

5. The Dalai Lama told this story and explained its implications at the Mind and Life XXIII meeting in Dharamsala, 2011. Daniel Goleman and John Dunne, eds., *Ecology, Ethics and Interdependence* (Boston: Wisdom Publications, 2017).

6. Anders Ericsson and Robert Pool, *Peak: Secrets from the New Science of Expertise* (New York: Houghton Mifflin Harcourt, 2016).

7. T. R. A. Kral et al., "Meditation Training Is Associated with Altered Amygdala Reactivity to Emotional Stimuli," under review, 2017.

8. J. Wielgosz et al., "Long-Term Mindfulness Training Is Associated with Reliable Differences in Resting Respiration Rate," *Scientific Reports* 6 (2016): 27533; doi:10.1038/srep27533.

9. Jon Kabat-Zinn et al., "The Relationship of Cognitive and Somatic Components of Anxiety to Patient Preference for Alternative Relaxation Techniques," *Mind/Body Medicine* 2 (1997): 101–9.

10. Richard Davidson and Cortland Dahl, "Varieties of Contemplative Practice," *JAMA Psychiatry* 74:2 (2017): 121; doi:10.1001/jamapsychiatry.2016.3469.

11. See, e.g., Daniel Goleman, *The Meditative Mind* (New York: Tarcher/Putnam, 1996; first published 1977 as *The Varieties of the Meditative Experience*). Dan now sees that categorization as limited in many ways. For one, this binary typing omits or otherwise conflates several important contemplative methods like visualization, where you generate an image and the set of feelings and attitudes that go with it.

12. Cortland J. Dahl, Antoine Lutz, and Richard J. Davidson, "Reconstructing and Deconstructing the Self: Cognitive Mechanisms in Meditation Practice," *Trends in Cognitive Science* 20 (2015): 1–9; http//dx.doi.org/10.1016/j.tics.2015.07.001.

13. Hazrat Ali, quoted in Thomas Cleary, *Living and Dying in Grace: Counsel of Hazrat Ali* (Boston: Shambhala, 1996).

14. Paraphrased from Martin Buber, *Tales of the Hasidim* (New York: Schocken Books, 1991), p. 107.

15. The Third Khamtrul Rinpoche, trans. Gerardo Abboud, *The Royal Seal of Mahamudra* (Boston: Shambhala, 2014).

16. J. K. Hamlin et al., "Social Evaluation by Preverbal Infants," *Nature* 450:7169 (2007): 557–59; doi:10.1038/nature06288.

17. F. Ferrarelli et al., "Experienced Mindfulness Meditators Exhibit Higher Parietal-Occipital EEG Gamma Activity during NREM Sleep," *PLoS One* 8:8 (2013): e73417; doi:10.1371/journal.pone.0073417.

18. The view that science and religion occupy different realms of authority and ways of knowing, and that these do not overlap, has been advocated, for example, by Stephen Jay Gould in *Rocks of Ages: Science and Religion in the Fullness of Life* (New York: Ballantine, 1999).

CHAPTER FOURTEEN. A HEALTHY MIND

1. L. Flook et al., "Promoting Prosocial Behavior and Self-Regulatory Skills in Preschool Children through a Mindfulness-Based Kindness Curriculum," *Developmental Psychology* 51:1 (2015): 44–51; doi:http://dx.doi.org/10.1037/a0038256.

2. R. Davidson et al., "Contemplative Practices and Mental Training: Prospects for American Education," *Child Development Perspectives* 6:2 (2012): 146–53; doi:10.1111/j.1750-8606.2012.00240.

3. Daniel Goleman and Peter Senge, *The Triple Focus: A New Approach to Education* (Northampton, MA: MoreThanSound Productions, 2014).

4. Daniel Rechstschaffen, *Mindful Education Workbook* (New York: W. W. Norton, 2016); Patricia Jennings, *Mindfulness for Teachers* (New York: W. W. Norton, 2015); R. Davidson et al., "Contemplative Practices and Mental Training: Prospects for American Education."

5. This work is still in its infancy and as of this writing, the first scientific articles assessing the games are being prepared for publication.

6. D. B. Levinson et al., "A Mind You Can Count On: Validating Breath Counting as a Behavioral Measure of Mindfulness," *Frontiers in Psychology* 5 (2014); http://journal.frontiersin.org/Journal/110196/abstract. Tenacity will likely be available in late 2017. For more info: http://centerhealthyminds.org/.

7. E. G. Patsenko et al., "Resting State (rs)-fMRI and Diffusion Tensor Imaging (DTI) Reveals Training Effects of a Meditation-Based Video Game on Left Fronto-Parietal Attentional Network in Adolescents," submitted 2017.

8. B. L. Alderman et al., "Mental and Physical (MAP) Training: Combining Meditation and Aerobic Exercise Reduces Depression and Rumination while Enhancing Synchronized Brain Activity," *Translational Psychiatry* 2 (accepted for publication 2016) e726–9; doi:10.1038/tp.2015.225.

9. Julieta Galante, "Loving-Kindness Meditation Effects on Well-Being and Altruism: A Mixed-Methods Online RCT," *Applied Psychology: Health and Well-Being* 8:3 (2016): 322–50; doi:10.1111/aphw.12074.

10. Sona Dimidjian et al., "Web-Based Mindfulness-Based Cognitive Therapy for Reducing Residual Depressive Symptoms: An Open Trial and Quasi-Experimental Comparison to Propensity Score Matched Controls," *Behaviour Research and Therapy* 63 (2014): 83–89; doi:10.1016/j.brat.2014.09.004.

11. Kathleen Garrison, "Effortless Awareness: Using Real Time Neurofeedback to Investigate Correlates of Posterior Cingulate Cortex Activity in Meditators' Self-Report," *Frontiers in Human Neuroscience* 7:440 (August 2013): 1–9.

12. Judson Brewer et al., "Mindfulness Training for Smoking Cessation: Results from a Randomized Controlled Trial," *Drug and Alcohol Dependence* 119 (2011b): 72–80.

13. A. P. Weible et al., "Rhythmic Brain Stimulation Reduces Anxiety-Related Behavior in a Mouse Model of Meditation Training," *Proceedings of the National Academy of Sciences*, in press, 2017. The photic driving impact of strobe lights can create a danger in humans for those with epilepsy, because the rhythms can sometimes trigger a seizure.

14. H. F. Iaccarino et al., "Gamma Frequency Entrainment Attenuates Amyloid Load and Modifies Microglia," *Nature* 540:7632 (2016): 230–35; doi:10.1038/nature20587.

15. The mouse's basic mammalian biology maps somewhat along human lines, but not entirely, and when it comes to the brain, the differences are far greater.

16. For more details, see Daniel Goleman, *A Force for Good: The Dalai Lama's Vision for Our World* (New York: Bantam, 2015); www.joinaforce4good.org.

17. Some evidence for this strategy: C. Lund et al., "Poverty and Mental Disorders: Breaking the Cycle in Low-Income and Middle-Income Countries," *Lancet* 378:9801 (2011): 1502–14; doi:10.1016/S0140-6736(11)60754-X.

Index

INDEX

mind-wandering, 151, 152, 251, 252
Mingyur Rinpoche, 216–21, 223, 224, 226, 228, 239, 262
monkey mind, 36, 156
mood, 185
moral guidelines, 270
multitasking, 136–38
Munindra, Anagarika, 22–23, 35
musical training, 301n12

Namgyal Monastery Institute of Buddhist Studies, 244–45
Nass, Clifford, 138
National Center for Complementary and Integrative Health, 188
National Institute of Health, 175, 188, 282, 290
National Institute of Mental Health (NIMH), 198–99
Neem Karoli Baba, 20–22, 262
Neff, Kristin, 105
Negi, Geshe Lobsang Tenzin, 117–18, 317n8
neocortex, 127
neural profile, 109–10, 117
neurofeedback, 286–88
neurogenesis, 283
neuromythology, 184–87
neuroplasticity, 50–52, 252, 278–83, 302n16
neuroscience, 7–8, 29, 47–49, 127, 129, 265, 268. *See also* brain; research; research methods
neurotransmitters, 32, 299n17
Neville, Helen, 51–52, 301n14
New York University, 27–28, 298nn9–10
nibbana, 38, 216
Nisker, Wes, 297n3
nondual stance, 263, 264
nonfindings, 65, 143
Norla Lama, 262
Northeastern University, 116–17
nucleus accumbens, 158, 162–63, 247, 252, 254
null findings, 61
Nyanaponika Thera, 41, 43, 300n2
Nyanatiloka Thera, 300n2

open awareness, 135–36
open presence, 221, 236, 239

orbitofrontal cortex, 180
orienting, 129–30, 143
Oxford University, 195–97

pain, 147–49, 166–67, 238–41
and meditation, 88–91, 193, 195
neural networks for, 114
and yogis, 248, 253–54
panchakarma, 177–78
parafoveal vision, 51, 301nn13–14
paramitas, 266–67
patience, 257, 267
peak experiences, 299n15
peripheral vision, 51, 301nn13–14
personality, 302n16
photic driving, 287, 321n13
physiological measures, 29, 62
Pinger, Laura, 278–80
positive feelings, 114
Posner, Michael, 153
postcingulate cortex (PCC), 150–51, 156, 160, 189, 238, 285–86
post-traumatic stress disorder, 1–2, 52, 199–202
prefrontal cortex, 127–28, 179, 180
activation of, 185, 243
and amygdala, 97
and early meditation stages, 160
and long-term meditators, 252
and PCC, 238
primary auditory cortex, 302n14
Prison Mindfulness Institute, 277
psychedelics, 299n17
psychoanalysis, 205–6
psychological symptoms, 167–68, 202–4
psychology
behavioral, 27–28, 298n8
Buddhist, 41–42
clinical, 25
early days of, 31
psychology of consciousness course, 43, 205
psychotherapy, 191–207
purpose, 56, 57, 92–93, 270

questionnaires, 75–77

Raichle, Marcus, 149–50
Ram Dass, 21, 297n3
reactivity level, 97
refractory period, 134